Chemistry Research and Applications

www.novapublishers.com

Chemistry Research and Applications

Polycyclic Aromatic Hydrocarbons: Sources, Exposure and Health Effects
Warren L. Gregoire (Editor)
2022. ISBN: 978-1-68507-626-9 (Softcover)
2022. ISBN: 978-1-68507-685-6 (eBook)

Cyanide: Occurrence, Applications and Toxicity
Bill M. Torres (Editor)
2022. ISBN: 978-1-68507-619-1 (Softcover)
2022. ISBN: 978-1-68507-670-2 (eBook)

A Closer Look at Carvacrol
Zak A. Cunningham (Editor)
2022. ISBN: 978-1-68507-627-6 (Softcover)
2022. ISBN: 978-1-68507-634-4 (eBook)

Fundamentals of Photocatalysis
Orva Auger (Editor)
2021. ISBN: 978-1-68507-374-9 (Softcover)
2021. ISBN: 978-1-68507-417-3 (eBook)

Polypropylene: Advances in Research and Applications
Théodore Marleau (Editor)
2021. ISBN: 978-1-68507-378-7 (Hardcover)
2021. ISBN: 978-1-68507-401-2 (eBook)

Boron: Advances in Research and Applications
Lynn Mcconnell (Editor)
2021. ISBN: 978-1-68507-231-5 (Hardcover)
2021. ISBN: 978-1-68507-259-9 (eBook)

More information about this series can be found at
https://novapublishers.com/product-category/series/chemistry-research-and-applications/

Güllü Kaymak
Editor

The Science of Carbamates

DOI: https://doi.org/10.52305/MADK9545

Library of Congress Cataloging-in-Publication Data

ISBN: 978-1-68507-708-2

Published by Nova Science Publishers, Inc. † New York

Contents

Preface

This scientific book is about the toxic effects of carbamates, which have a functional carbonyl group, and how these effects can be eliminated. It is aimed to provide knowledge to all scientists related to this field. Although carbamates are generally considered to be pesticides, significant amounts of carbamates are used as medicine. This book contains eight chapters that examine carbamates from a variety of perspectives, including the latest research in this field. Chapter One comprehensively studies the chemical structure of carbamates used as drugs and the poisoning of carbamates. Chapter Two discusses how to develop cheap and efficient modes of remediation from the environment. Chapter Three highlights the important role of microorganisms in the degradation of carbamates. Chapter Four focuses on biodeterioration of carbamates by fungi. Chapter Five examines the probiotic microorganisms as detoxification tools in the degradation and transformation of carbamates. Chapter Six reviews the genetic risk that may occur as a result of exposure to carbamate pesticides in living things. Chapter Seven explains the genetic biomarkers that help determine the management of chronic pesticide toxicity by determining different gene polymorphisms in agricultural workers as a result of long-term contact with carbamates. Chapter Eight thoroughly studies the clinical information about intoxication that may arise as a result of carbamate group exposure.

Acknowledgments

I would like to thank the lecturers of Kütahya Health Sciences University, Dear Melis Kutlu-Girgin, Dear Mehmet Gökçe and Dear Levent Sevi, who have been very helpful in the English editing and proofreading of the chapters.

I would like to acknowledge the publisher and support staff at Nova Science Publishers; they were very helpful in the various stages of producing this book.

Chapter 1

The Chemical and Toxicological Structure of Carbamates

Elif Aydın*
Disinfection, Sterilization and Antisepsis Program, Kütahya Health Sciences University, Kütahya, Turkey

Introduction

Pesticides are compounds formed with the combination of one or more substances and used to control, destroy and prevent pests such as insects, weeds, birds, rodents, nematodes, and microorganisms and have the property of damaging plants [1].

According to the definition of the FAO International Code of Conduct on the Distribution and Use of Pesticides, a pesticide is a substance or combination of substances intended for the prevention, eradication, and control ventors, undesirable insect species, and pests that cause disease in humans or animals. The objective of pesticide treatment is to protect crops such as fruits and vegetables during planting, growth, harvest storage, and transportation, i.e., before and after harvest [2].

Pesticides are classified according to their chemical, physical structure, and target species, inorganic, synthetic, and biological. Although their distinction is not obvious, they can also be classified as inorganic, synthetic, and biological (biopesticides) pesticides. Biopesticides include microbial and biochemical pesticides [3]. Years ago, various natural substances were used as pesticides. Then sulfur, metal salts, tobacco products, and natural oils were

* Corresponding Author's Email: elif.aydin@ksbu.edu.tr.

In: The Science of Carbamates
Editor: Güllü Kaymak
ISBN: 978-1-68507-708-2

used. Since the 1970s, the chemical synthesis of pesticides has increased, and today there are more than 55 classes and 1500 individual substances with more than 100,000 formulas.

Pesticides are classified into the following groups;

- Insecticides
- Other insect control agents

    - Chemosterilants
    - Pheromones (sexual attractants and synthetic baits)
    - Repellents
    - Insect hormones and hormone mimics (insect growth regulators)
    - Herbicides

- Protective fungicides
- Plant growth regulators
- Rodenticides
- Soil fumigants and nematocides
- Eradican fungicides (chemotherapy drugs)
- Desiccants, defoliants and sap killers
- Specific acaricides
- Molluscicides

The four major pesticide groups are carbamates, organophosphates, organochlorines, and pyrethroids [4]. Carbamates and organophosphates have replaced organochlorines due to their corrected and improved effects. 200 different types of organophosphorus esters and 25 types of carbamic acid esters are available in the world market [5].

Since carbamate compounds are considered to be less persistent and have less toxicity than the neurotoxic organophosphates (OPs) compared to organochlorines (OCs), they and related compounds have replaced organophosphates (OPs). Carbamates are chemical ester compounds derived from carbamic acids. They are structurally simpler and are considered safer than OPs. Due to their easier degradation, low bioaccumulation, and lower persistence, they are widely used in agriculture as well as in households and industrial workplaces. Due to their broad spectrum of biological horizons, Carbamates are widely used in modern agriculture as insecticides, herbicides, fungicides, nematicides, molluscicides, and acaricides. In the last two decades, the widespread use of carbamate compounds in pest control has increased more than that of OCs or OPs.

Carbamic acid: It is an unstable carbamate compound. It is technically the simplest amino acid but is unstable due to the nature of the carboxyl-nitrogen bond, so the simplest amino acid is not carbamic acid but glycine. It is important for naming large molecules. Carbamic acid cannot be synthesized or characterized by any experimental technique. Carbamic acids are intermediates released during the degradation of carbamate protecting groups. Hydrolysis of an ester bond produces carbamic acid, which reverses the protecting reaction to release carbon dioxide and produce an unprotected free amine.

Carbamate is an ester of carbamic acid. Methyl carbamate is the simplest ester of carbamic acid. Some are used as insecticides, while others are used as muscle relaxants [6].

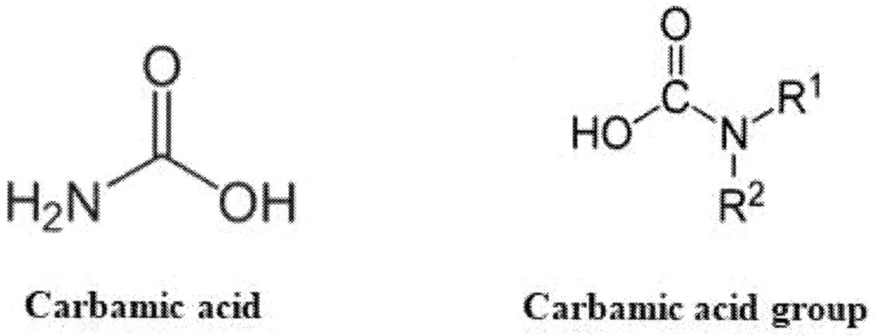

Carbamic acid **Carbamic acid group**

Figure 1. The chemical structures of carbamic acid and carbamic acid group.

Carbamates

Carbamates are compounds derived from carbamic acid (NH_2COOH). A carbamate group, carbamate esters, and carbamic acids contain functional groups that are similar and often chemically interchangeable. Carbamate esters are also known as urethanes. Carbamates are drug-active compounds that can be synthesized using alcohol and carbamoyl chloride. Carbamates are stable crystalline compounds commonly used as recognition reagents for alcohols [7].

As seen in Figure 2, the chemical compositions of carbamates contain three different functional units, R1, R2, and R3. In general, R1 and R2 represent organic functional groups.

When R1 is methyl and R2 is a hydrogen atom, this confers insecticidal properties to the carbamate when R1 has an aromatic functional group, it is a herbicide, and when R1 is a benzimidazole moiety, it confers fungicidal properties.

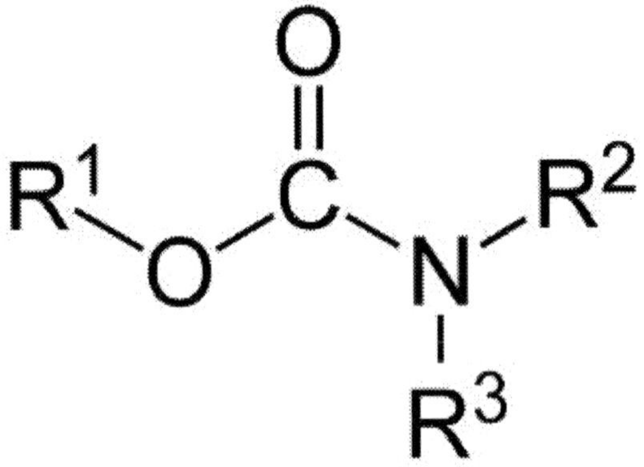

Figure 2. Chemical structure of carbamates.

Synthesis of carbamate derivatives: Carbamic acids are derived from amines.

$R_2NH + CO_2 \rightarrow R_2NCO_2H$

Carbamic acid is almost as acidic as acetic acid. Ionization of a proton yields a carbamate an ion which is conjugated to carbamic acid.

$R_2NCO_2H \rightarrow R_2NCO_2^- + H^+$

Carbamates are also formed by the hydrolysis of chloroformamides.

$R_2NC(O)Cl + H_2O \rightarrow R_2NCO_2H + HCl$

Carbamates are also formed by Curtius translation, where the isocyanates are formed to interact with alcohol.

$RNCO + R'OH \rightarrow RNHCO_2R'$

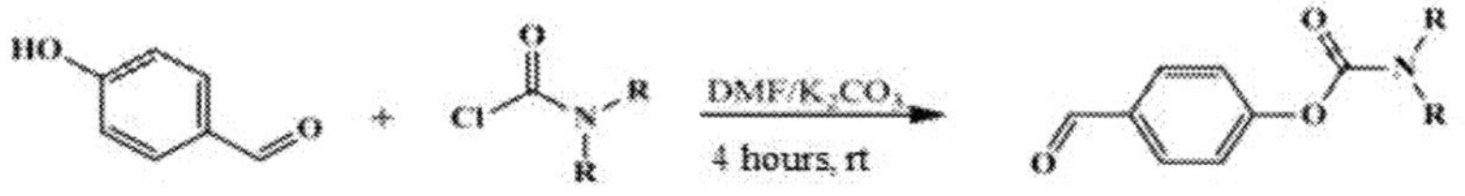

Figure 3. General synthesis reaction of carbamate derivatives.

Unlike OPs, carbamates are reversible acetylcholinesterase (AChE) inhibitors, but the mechanisms of toxicity are the same. Acetylcholine (ACh) is the enzyme required for the proper transmission of nerve signals to tissues, skeletal muscles, and the brain. Inhibition of AChE disrupts the function of the neurotransmitter acetylcholine at synapses in the brain, resulting in ACh accumulation and toxicity.

Carbamates, inactive derivatives of the drug in the pharmaceutical industry, can be used as an anesthetic in animals and as an insecticide in agriculture [8]. Biodegraded carbamates are carbamates used as acetylcholinesterase inhibitors, anticytotoxic, antimicrobial, antifungal, antibacterial, and acetylcholinesterase inhibitors. In addition, surgical drapes and dressings used in the medical industry are made of polyurethane, a carbamate derivative. Urethane or ethyl carbamate has been produced commercially in the USA for cancer treatment and other medical purposes. It is also sometimes used in veterinary medicine. In addition, some carbamates are also used in human treatment [9, 10].

Neostigmine: Neostigmine, an acetylcholinesterase inhibitor synthesized by Aeschlimann and Reinert in 1931 and containing carbamate in its structure, is the first synthetic drug that can be used to treat Myasthenia Gravis [11].

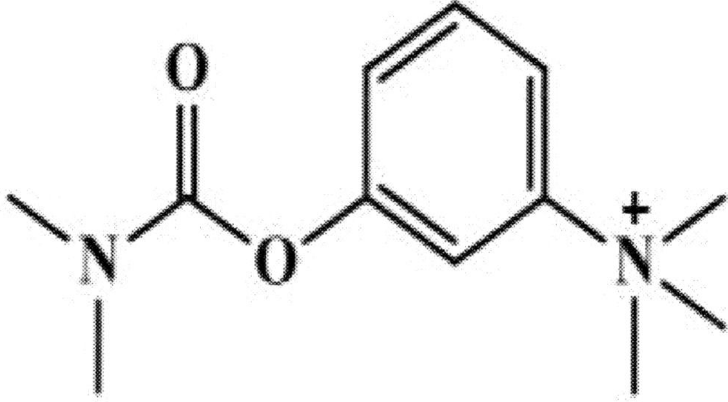

Figure 4. The chemical structure of neostigmine.

Some of the muscles that move voluntarily, especially the muscles that move the eye, allow chewing and swallowing, exhaust quickly, and weaknesses of these muscles are the symptoms of Myasthenia gravis. These symptoms due to constant movements disappear with the administration of neostigmine, an anticholinesterase inhibitor.

Rivastigmine: Rivastigmine is a carbamate derivative approved by the FDA in 1999, and since then has been used in the treatment of Alzheimer's disease. Rivastigmine is a central nervous system-selective, an irreversible acetylcholinesterase inhibitor [12, 13].

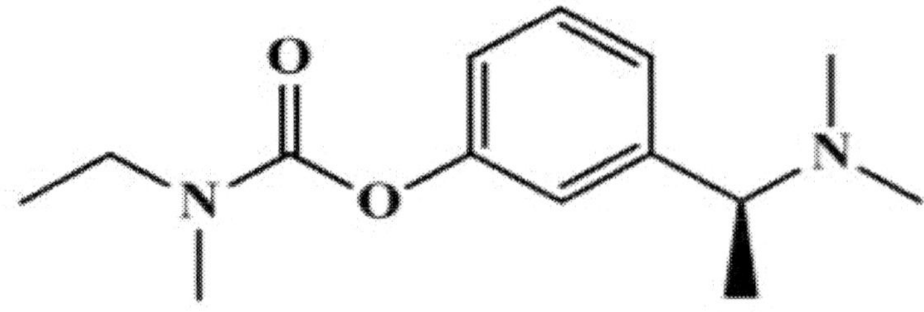

Figure 5. Chemical structure of rivastigmine.

Sulfur Analogs: A carbamate contains two oxygen atoms, ROC (= O)NR2, one or both of which may be replaced by sulfur. Carbamate analogs in which only one of the oxygen atoms are replaced by sulfur are called thiocarbamate.

Carbamates in which both oxygens are replaced by sulfur are called dithiocarbamates and are referred to as RSC (= S)NR2.

There are two different structural isomeric types of thiocarbamate:

O-thiocarbamates, ROC (= S)NR2, in which the carbonyl group (C=O) is replaced by a thiocarbonyl group (C = S).

S-thiocarbamates, RSC (= O)NR2, in which the R-O- group is replaced by the R-S- group. O-thiocarbamates can isomerize to S-thiocarbamates [14].

Methyl Carbamate: Methyl carbamate (methylurethane or generated) is an organic compound. It is the simplest ester of the fictitious carbamic acid (NH_2COOH).

Methyl carbamate is formed by the reaction of ammonia with methyl chloroformate or methyl carbonate. Besides boron trifluoride, methanol can also interact with urea. Unlike ethyl carbamate, it is mutagenic in Salmonella but not mutagenic in Drosophila. Experimental evidence has shown it to be carcinogenic to the rats but not to the mice.

Methyl carbamate is used in the textile industry to produce resins that are used as a pressure-resistant finishing agent in polyester/cotton blended fabrics. N-Methyl carbamates are widely used as insecticides [15].

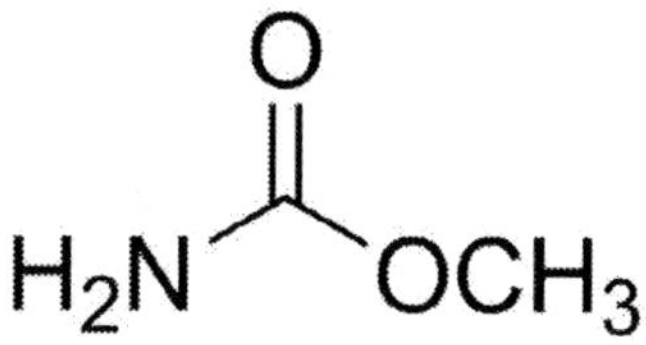

Figure 6. Chemical structure of methyl carbamate.

Mebendazole: Mebendazole is a broad spectrum benzimidazole drug. It is an anthelmintic, used in the treatment against worms and as adjunctive therapy in before and after surgical procedures. Recently, mebendazole has been used as an anti-cancer medicine [16].

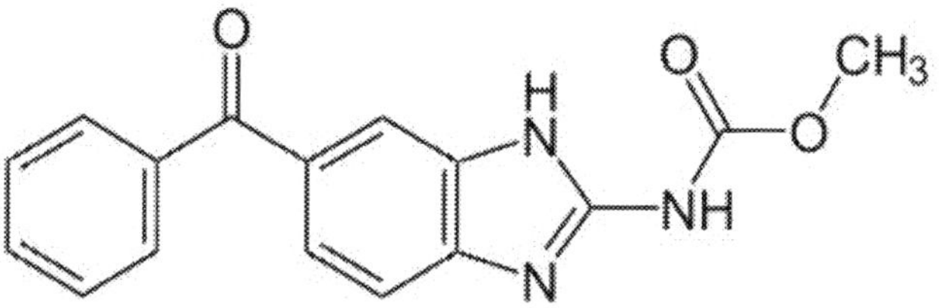

Figure 7. Chemical structure of mebendazole.

Ethyl Carbamate: Ethyl carbamate (urethane) is an ester of carbamic acid. Despite its common name, it is not a component of polyurethanes. Ethyl carbamate is a white crystalline substance obtained by the action of ammonia on ethyl chloroformate or by heating urea nitrate and ethyl alcohol. Ethyl carbamate is used as an anti-cancer medicine and for other medical purposes [17].

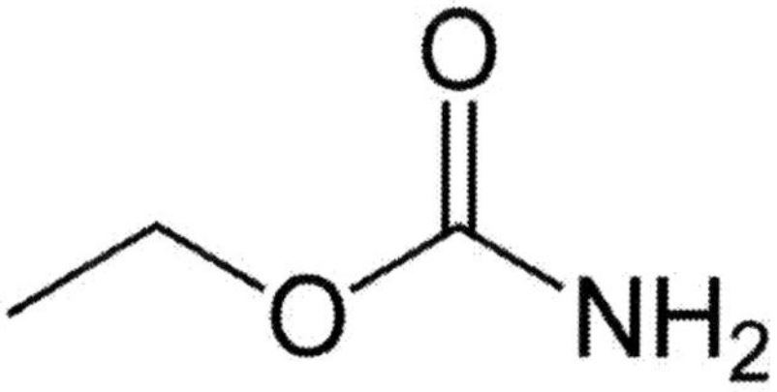

Figure 8. Chemical structure of ethyl carbamate.

Albendazole: Albendazole is also known as albenza eskazol, zentel and andazol. It is a potent, broad-spectrum anthelmintic agent used in medicine and veterinary applications and is a member of benzimidazole compounds. In addition to its anthelmintic properties, Albendazole is used as an anti-cancer medicine [18].

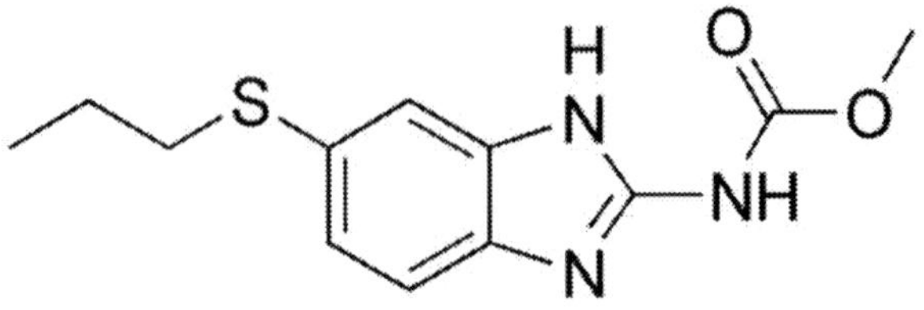

Figure 9. Chemical structure of albendazole.

Asulam: Asulam is a herbicide developed by Bayer. It is used in agriculture and horticulture. It is used as an antiviral agent as well as to kill ferns and weeds [19].

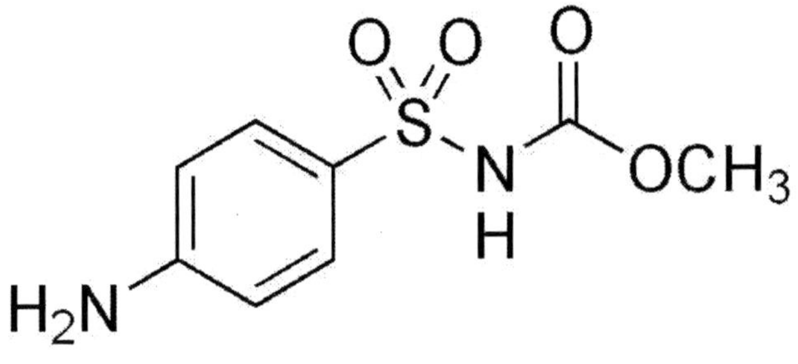

Figure 10. Chemical structure of Asulam.

Benomyl: Benomyl is a fungicide and inhibits fungal growth on a wide variety of plants. It is a systemic benzimidazole fungicide that is selectively poisonous to microorganisms, invertebrates, and especially earthworms [17].

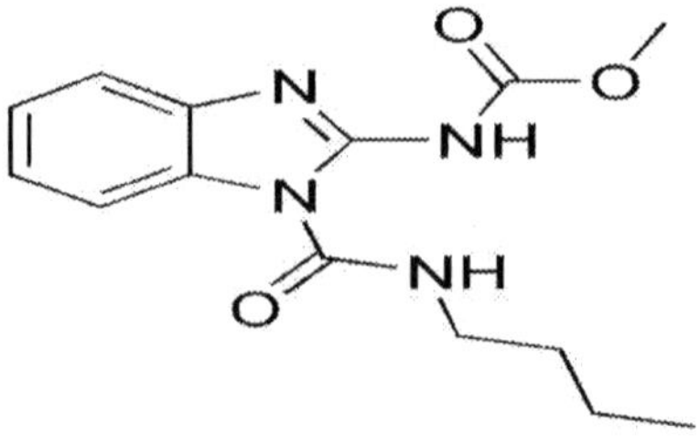

Figure 11. Chemical structure of benomyl.

Carbendazim: Carbendazim is a widely used systemic benzimidazole fungicide with a broad spectrum of activity. It is also mainly used as a worm control agent in turf situations such as tennis courts, golf greens, etc., and is registered in some countries for this use only [17].

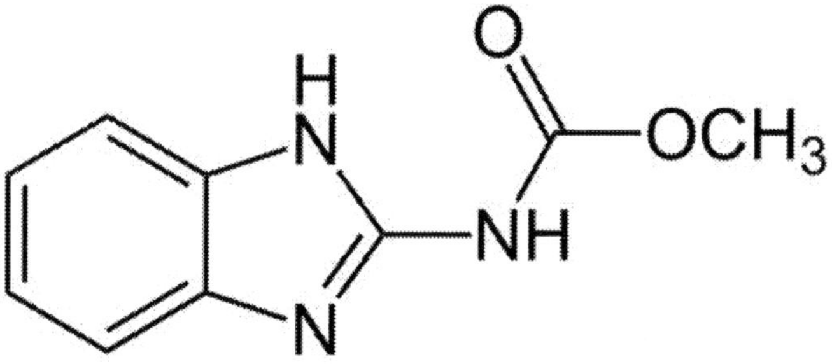

Figure 12. Chemical structure of carbendazim.

Cyclobendazole: Cyclobendazole is an anthelmintic. Cyclobendazole is insoluble in water and organic solvents [17].

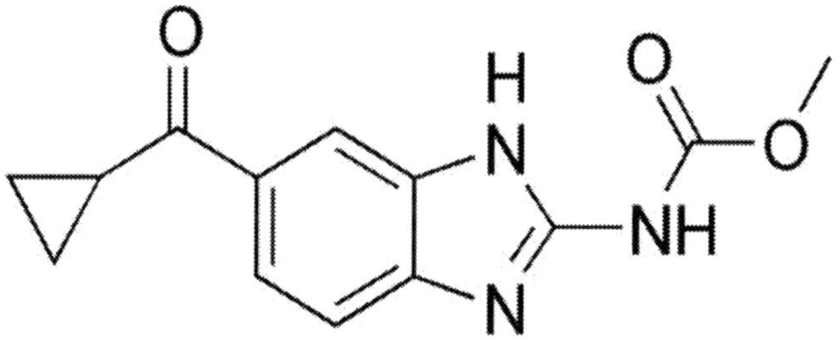

Figure 13. Chemical structure of cyclobendazole.

Cloforex: Cloforex (obereks) is an appetite suppressant in the amphetamine class [17].

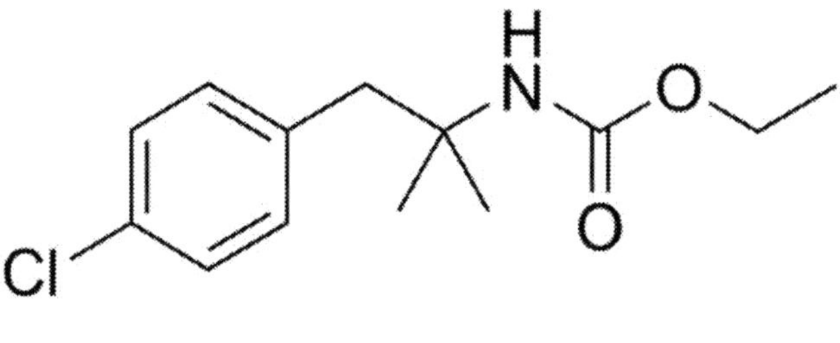

Figure 14. Chemical structure of chloforex.

Fenbendazole: Fenbendazole is a broad-spectrum anthelmintic used against gastrointestinal parasites such as giardia, roundworms, hookworms, some tapeworms, pinworms, etc. It can be used as a hydra treatment in sheep, cattle, horses, fish, cats, dogs, rabbits, many reptiles, shrimps, and seals [20].

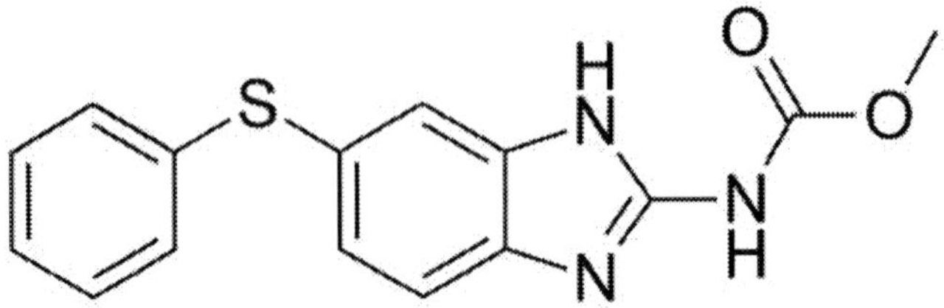

Figure 15. Chemical structure of fenbendazole.

Flubendazole: Flubendazole is an anthelmintic. Its trade name is flutelmium. It is used by veterinarians as a preservative against internal parasites and worms in cats and dogs. Other trade names such as Flubenol, Bioovermin and Flumoxal are also used [21].

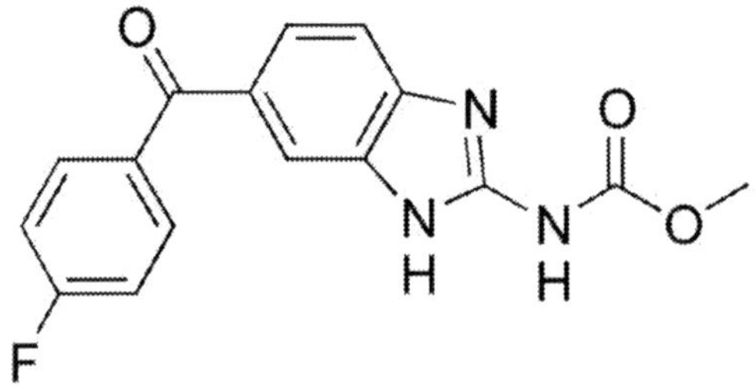

Figure 16. Chemical structure of flubendazole.

Flupirtine: Flupirtine was originally used as an analgesic for acute and chronic pain, but in 2013, the European Medicines Agency restricted its use to acute pain only due to liver toxicity. It was reported that it should not be used for more than two weeks in patients with pain. In March 2018, the marketing authorization for flupirtine was withdrawn on the recommendation of the European Medicines Agency because the restrictions introduced in 2013 were not adequately followed in clinical practice and cases of severe liver injury, including liver failure, continued to occur [22].

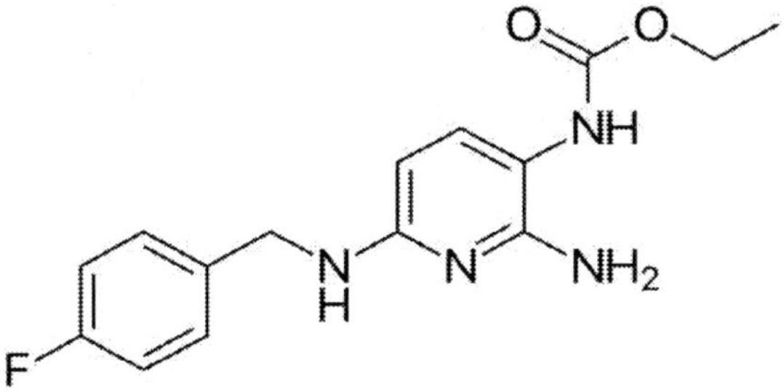

Figure 17. Chemical structure of flupirtine.

Moricizine: Moricizine is a phenothiazine derivative with antiarrhythmic properties [23].

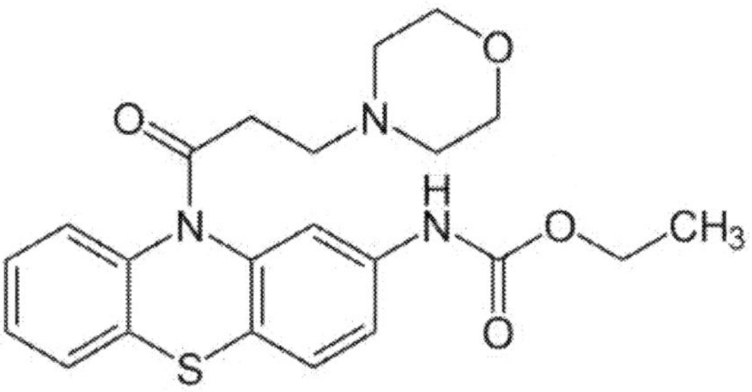

Figure 18. Chemical structure of moricizine.

Nocodazole: Microtubules are a type of fibers that form the cytoskeleton. The dynamic microtubule network has many important functions in the cell, including vesicular transport, mitotic spindle formation, and cytokinesis. Nocodazole is an antineoplastic agent that acts in cells by interfering with the polymerization of microtubules. Nocodazole is also used as anti-cancer medicine [24].

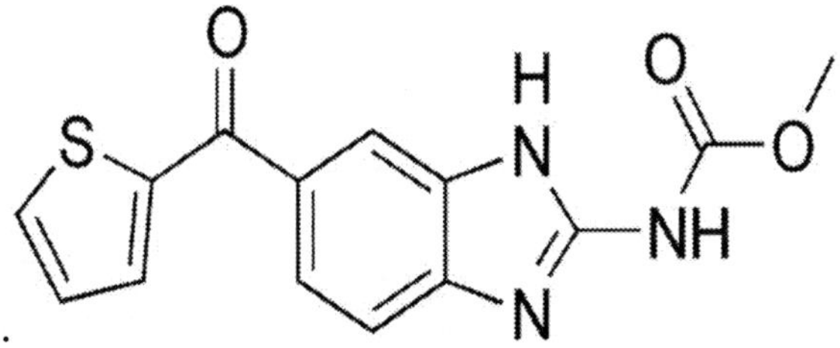

Figure 19. Chemical structure of nocodazole.

Oxfendazole: Oxfendazole is a Broad-spectrum anthelmintic that primarily protects livestock against roundworms, suck-worms, and pinworms. This drug is rarely used in horses, goats, sheep, and cattle. It is used very little in dogs and cats. The drug for farm animals is largely available in oral dosage forms.

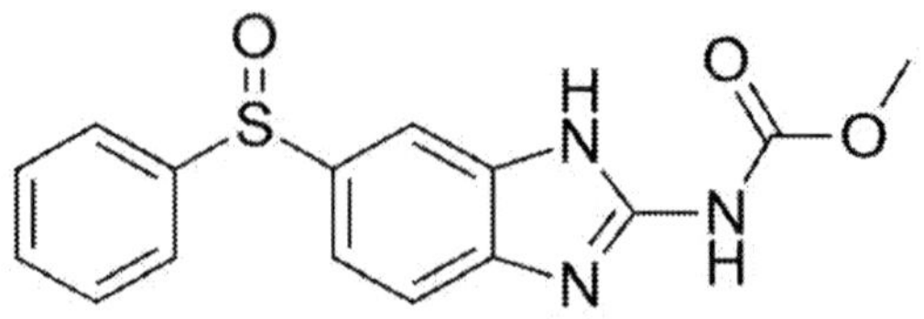

Figure 20. Chemical structure of oxfendazole.

Retigabine: Retigabine (ezogabine) is an anticonvulsant used as adjunctive therapy for partial epilepsy in adult patients. Retigabine was developed by Valeant Pharmaceuticals and Glaxo Smith Kline. It was placed on the market by the European Medicines Agency on March 28, 2011, and was discontinued in June 2017.

Retigabine acts as a potassium channel opener by activating a specific family of voltage-gated potassium channels in the brain. It is an antiepileptic drug and promising for treating other neurological conditions, including tinnitus, migraine, and neuropathic pain [26].

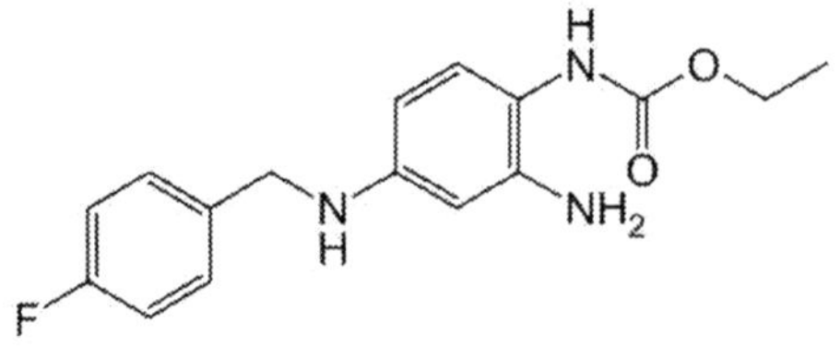

Figure 21. Chemical structure of retigabine.

Carbamate insecticides: Carbamate insecticides contain a carbamate ester group. This group of compounds includes fenoxycarb, aldicarb, carbaryl, carbofuran (Furadan), fenobucarb, and methiocarb. These insecticides kill insects by reversibly inactivating the enzyme acetylcholinesterase. The insect repellent icaridin is a substituted carbamate. In addition, some carbamates whose chemical structure is related to the natural alkaloid physostigmine are used in human therapy. Examples include the cholinesterase inhibitors neostigmine and rivastigmine. Other examples are meprobamate and its derivatives such as carisoprodol, felbamate, and tybamate, which include a class of anxiolytics and muscle relaxants and are widely used [27].

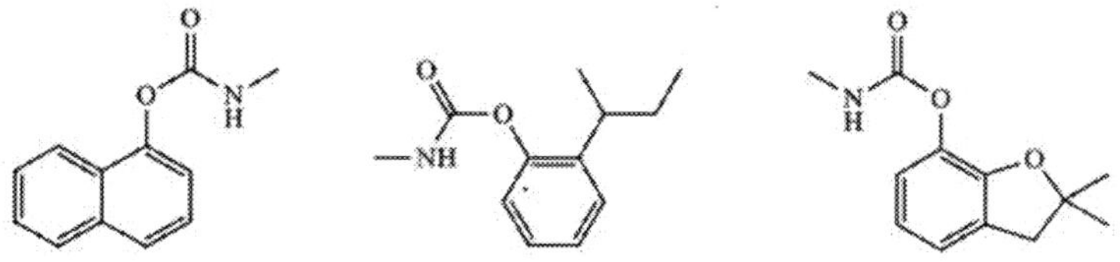

Carbofuran Fenobucarb Carbaryl

Figure 22. Chemical structures of carbofuran, fenobucarb and carbaryl.

Carbamate poisoning: Carbamates are absorbed orally or through the skin, and can cause serious health problems ranging from throat and eye infections, infections to skin irritation, skin irritation, coughing, and death. Carbofuran, carbaryl, aldicarb, carbendazim, and methomyl are recognized by the World Health Organization as endocrine disrupting carbamates. Carbamate compounds cause toxic, cytotoxic, genotoxic developmental and behavioral disorders. In addition, carbaryl and carbofuran cause reproductive problems in humans and animals such as decreased sperm production, altered spermatogenesis, and impaired testosterone levels. Carbendazim, on the other hand, causes developmental toxicity by decreasing progesterone and estradiol levels [27].

The main signs and symptoms that occur in poisoning with carbamate insecticides also involve the muscarinic and nicotinic cholinergic systems, the central nervous system, the respiratory system, and the cardiovascular system [5]. The signs and symptoms can occur within a few minutes to 12 hours. AChE enzyme activities in both erythrocytes and plasma are suppressed. The severity of clinical signs and symptoms parallels the suppression of AChE activity. The cause of death in carbamate insecticide poisonings is usually respiratory arrest due to paralysis of the respiratory muscles [28].

The diagnosis of poisoning with carbamate insecticides is usually made based on history or clinical suspicion, the presence of poisoning findings, and cholinesterase levels. In all patients suspected of having been exposed to carbamate insecticide, the time of exposure to the substance, the accompanying circumstances, the route of ingestion, the type and amount of carbamate insecticide should be questioned [5]. Biochemical tests such as electrolyte levels, pancreatic amylase, and urinalysis can also be used to evaluate the clinical condition of the patient in carbamate insecticide poisoning. ECG is one of the tests that must be performed. The incidence of respiratory failure and the risk of poor prognosis are increased in patients who develop QTc prolongation or PVC [29, 30].

Since poisoning with carbamate insecticides may result in severe, life-threatening clinical symptoms within a few hours of exposure, life-saving emergency and supportive treatment should be given immediately and specific antidote therapy should be initiated without delay. In this regard, treatment should be given in three basic steps [5].

- i *Emergency and supportive treatment:* It consists of providing and maintaining the patency of the airway, providing and supporting the circulation [31].
- ii *Prevention of absorption:* If contamination occurs through the skin, the patient's clothes should be removed and the skin should be washed with soap and warm water. In the first hour after oral ingestion, if the patient is conscious and not vomiting, activated charcoal is administered and the stomach is washed [32).
- iii *Specific antidote therapy:* It includes the administration of atropine, and is a muscarinic receptor antagonist, and oximes (pralidoxime) which release the AChE enzyme. The duration and amount of the subsequent dose are determined according to the patient's condition [5].

Conclusion

A carbamate is a category of organic compounds that is formally derived from carbamic acid (NH_2COOH). The term includes organic compounds (e.g., the ester ethyl carbamate), formally obtained by replacing one or more of the hydrogen atoms by other organic functional groups; as well as salts with the carbamate anion H_2NCOO^-. Within nature carbon dioxide can bind with neutral amine groups to form a carbamate, this post-translational modification is known as carbamylation. This modification is known to occur on several important proteins. In this chapter, important carbamate compounds and their poisonings are mentioned in general.

References

[1] Umar Mustapha M, Halimoon N, Lutfi W, Johari W, Yunus M, Shukor A, Halimoon N, Johar W, Yunus M. (2019). An Overview on Biodegradation of Carbamate Pesticides by Soil Bacteria. *Pertanika J Sc. and Technol*, 27, 547-563.

[2] Joint FAO/WHO Expert Committee on Food Additives, Joint FAO/WHO Expert Committee on Food Additives. Meeting, and World Health Organization. (2002). Evaluation of Certain Food Additives and Contaminants: Fifty-seventh Report of the Joint FAO/WHO Expert Committee on Food Additives (Vol. 57). World Health Organization.

[3] Thakur N, Kaur S, Tomar P, Thakur S and Yadav AN. (2020). Microbial biopesticides: Current status and advancement for sustainable agriculture and environment. In *New and Future Developments in Microbial Biotechnology and Bioengineering*, 243-282.

[4] Kaymak, G. (2020). Experimental evidence of the sublethal effects of carbamate pesticide on zebrafish (*Danio rerio*). *Fresenius Environmental Bulletin*, 29(9), 7267-7274.

[5] Tuncok Y, Aksay Hocaoğlu N. (2006). Organofosfatlı İnsektisidlerle Zehirlenme, *Turkiye Klinikleri J Surg Med Sci,* 2: 69-73.

[6] Matošević A and Bosak A. (2020). Carbamate group as structural motif in drugs: A review of carbamate derivatives used as therapeutic agents. *Archives of Industrial Hygiene and Toxicology*, 71(4), 285.

[7] Solomons TWG, Fryhle CB, Wiley J and Sons N. (2002). Radical reactions. *Organic Chemistry*. 7th ed, Wiley, New York, 146-211.

[8] Liu JF, Sang CY, Xu XH, Zhang LL,Yang X, Hui L, Zhang JB, Chen SW, (2013). Synthesis and cytotoxic activity on human cancer cells of carbamate derivatives of 4b-(1,2,3- triazol-1-yl)podophyllotoxin, *European Journal of Medicinal Chemistry*, 64, s.621-628.

[9] Kratky M, Volkova M, Novotna E, Trejtnar F, Stolarikova J, Vinšova J, (2014). Synthesis and biological activity of new salicylanilide N,N-disubstituted carbamates and thiocarbamates, *Bioorganic and Medicinal Chemistry,* 22, s.4073-4082.

[10] Mustazza C, Borioni A, Giudice MRD, Gatta F, Ferretti R, Meneguz A, Volpe MT, Lorenzini P. (2002). Synthesis and cholinesterase activity of phenylcarbamates related to Rivastigmine, a therapeutic agent for Alzheimer's disease, *Eur J Med Chem*, 37, s.91-109.

[11] Porst H, Kny L. (1985). The structure of degradation products of neostigmine bromide, *Die Pharmazie*, 40, s.325-328.

[12] Rösler M, Anand R, Cicin-Sain A, Gauthier S, Agid Y, Dal-Bianco P, Stahelin HB. (1999). Efficacy and Safety of Rivastigmine in Patients with

Alzheimer's disease, *International Randomised Controlled-Trial,* 318, s.633-640.

[13] Corey-Bloom JP. (1998). A randomized trial evaluating the efficacy and safety of ENA 713 (rivastigmine tartrate), a new acetylcholinesterase inhibitor, in patients with mild to moderately severe Alzheimer's disease. *Int J Geriatr Psyopharmacol*, 1, 55-65.

[14] International Programme on Chemical Safety and World Health Organization. (1986). *Carbamate pesticides: A general introduction.* World Health Organization. https://apps.who.int/iris/handle/10665/38687.

[15] Foureman P, Mason JM, Valencia R and Zimmering S. (1994). Chemical mutagenesis testing in Drosophila. X. Results of 70 coded chemicals tested for the national toxicology-program. *Environmental and molecular mutagenesis*, 23(3), 208-227.

[16] Sasaki JI, Ramesh R, Chada S, Gomyo Y, Roth JA and Mukhopadhyay T. (2002). The Anthelmintic Drug Mebendazole Induces Mitotic Arrest and Apoptosis by Depolymerizing Tubulin in Non-Small Cell Lung Cancer Cells 1 Supported in part by grants from the National Cancer Institute and the NIH Specialized Program of Research Excellence in Lung Cancer P-50-CA70907 and P01 CA78778-01A1 (both to JAR), by gifts to the Division of Surgery and Anesthesiology from Tenneco and Exxon for the Core Laboratory Facility, by The University of Texas MD Anderson Cancer Center Support Core Grant CA16672by the W. M. Keck Foundation, and by a sponsored research agreement with Introgen Therapeutics, Inc. J. A. R. is a scientific advisor for Introgen Therapeutics, Inc. *Molecular Cancer Therapeutics,* 1(13), 1201-1209.

[17] Budavari S, O'Neil MJ, Smith A and Heckelman PE. (1989). The Merck Index (Vol. 11, pp. 2330-2331). Rahway, NJ: Merck.

[18] Tamarozzi F, Horton J, Muhtarov M, Ramharter M, Siles-Lucas M, Gruener B, ... and Brunetti E. (2020). A case for adoption of continuous albendazole treatment regimen for human echinococcal infections. *PLoS Neglected Tropical Diseases,* 14(9), e0008566.

[19] Pakeman RJ, Le Duc MG and Marrs RH. (1998). An assessment of aerially applied asulam as a method of long-term bracken control. *Journal of Environmental Management,* 53(3), 255-262.

[20] Düwel D. (1977). Fenbendazole. II. Biological properties and activity. *Pesticide Science*, 8(5), 550-555.

[21] Khachigian LM. (2021). Emerging insights on the functions of the anthelmintic flubendazole as a repurposed anticancer agent. *Cancer Letters.*

[22] Reike-Kunze M, Zenouzi R, Hartel J, Krech T, Weidemann S, Sterneck M ... and Sebode M. (2021). Drug-induced liver injury at a tertiary care centre

in Germany: Model for end-stage liver disease is the best predictor of outcome. *Liver International,* 41(10), 2383-2395.

[23] Han C, Wirianto M, Kim E, Burish MJ, Yoo SH and Chen Z. (2021). Clock-Modulating Activities of the Anti-Arrhythmic Drug Moricizine. *Clocks and Sleep,* 3(3), 351-365.

[24] Vasquez RJ, Howell B, Yvon AM, Wadsworth P and Cassimeris L. (1997). Nanomolar concentrations of nocodazole alter microtubule dynamic instability in vivo and in vitro. *Molecular biology of the cell*, 8(6), 973-985.

[25] Gonzalez AE, Gavidia C, Falcon N, Bernal T, Verastegui M and Garcia HH. (2010). Protection of pigs with cysticercosis from further infections after treatment with oxfendazole.

[26] Nissenkorn A, Kornilov P, Peretz A, Blumkin L, Heimer G, Ben-Zeev B and Attali B. (2021). Personalized treatment with retigabine for pharmacoresistant epilepsy arising from a pathogenic variant in the KCNQ2 selectivity filter. *Epileptic Disorders*, 1(1).

[27] Karakuş M. (2014). *Alkoksikarbonilaminobenzoik asit esterlerinin sentezi* (Master's thesis, İnönü Üniversitesi Fen Bilimleri Enstitüsü).

[28] Richard F. Clark. Insecticides: Organophosporus compounds and carbamates. In: Goldfrank LR, Flomenbaum NE, Lewin NA, Weisman RS, Howland MA, Hoffman RS, eds. *Goldfrank's Toxicologic Emergencies*, 7th edition, USA, The McGraw-Hill Companies 2002; 1346-1378.

[29] Bardin PG, Van Eeden SF, Moolman JA, Foden AP and Joubert JR. (1994). Organophosphate and carbamate poisoning. *Archives of Internal Medicine,* 154(13), 1433-1441.

[30] Leibson T and Lifshitz M. (2008). Organophosphate and carbamate poisoning: Review of the current literature and summary of clinical and laboratory experience in southern Israel. *The Israel Medical Association Journal,* 10(11), 767.

[31] Zavon MR. (1964). Diagnosis and treatment of pesticide poisoning. Archives of Environmental Health: *An International Journal*, 9(5), 615-620.

[32] Goel A and Aggarwal P. (2007). Pesticide poisoning. *National Medical Journal of India,* 20(4), 182.

Chapter 2

Remediation of Carbamate Toxicity

Rahul Badru* **and Khushboo Verma**
Department of Chemistry, Sri Guru Granth Sahib World University, Fatehgarh Sahib, Punjab, India

Introduction

Carbamates, which are presently of extraordinary importance in the agriculture industry, were first prepared by the Geigy Corporation in 1947 [1]. Carbamates were first introduced in 1956 as an alternative to chlorine and phosphorus-based organic insecticides for pest management [2]. It was in 1959 when one of the most common carbamate pesticides propoxur (2-isopropoxyphenyl-*N*-methylcarbamate) was commercialized under the name of baygon (brand name Raid), which is still in use today. Another category of carbamates with the ester functionality includes aldicarb, carbofuran, carbaryl, ethienocarb, fenobucarb, oxamyl, and methomyl. Most of these are used as insecticides and the mode of action of these carbamates is to kill insects by inhibiting the acetylcholinesterase enzyme reversibly [3]. But fenoxycarb pesticide registered in 1985 did not show the same mode of action as that reported for other carbamates, rather it acts as a mimic for juvenile hormones in insects to prevent them to reach maturity [4]. Icaridin is another carbamate (also termed picaridin) employed as an insect repellent. In addition to these, the carbamates had been reported to show benefits like short-term and low mammalian toxicity, more biodegradability, and easy decomposition over other pesticides used before them [5].

* Corresponding Author's E-mail: rahulbadru@gmail.com.

In: The Science of Carbamates
Editor: Güllü Kaymak
ISBN: 978-1-68507-708-2

What Are Carbamates?

The carbamates are esters of an unstable carbamic acid (NH_2COOH), which upon substitution with alkyl or aryl groups at both amino and acid ends gives stable compounds.

In nature, carbamates are formed by carbonylation of amines with atmospheric CO_2, termed as carbamoylation. Some important proteins in nature, such as haemoglobin (Hb), have this form of alteration. The free amine groups of valine residue present in the α and β chains of deoxyhaemoglobin occur in the carbamate form. They just help in the stabilization of the protein in deoxy-form and thereby facilitate the release of oxygen, which is bound to the protein [6]. Similarly, the amine groups of the lysine amino-acid residues of enzymes like urease and phosphotriesterase do occur in the carbamate form. Few carbamates also play a role in the synthesis of biomolecules like carbamate derivative of 2-aminoimidazole serves as a precursor for the biogenesis of inosine. The biogenesis of carbamoyl phosphate also involves *N*-acetylglutamate which contains a carbamate moiety [7].

The carbamates are also formed during CO_2 uptake by plants. The enzyme involved in this natural fixation of CO_2 is ribulose-1,5-biphosphate carboxylase, which fixes atmospheric CO_2 into phosphoglycerate in the course of the Calvin-Benson Cycle. The CO_2 acts not merely as a substrate for carboxylation, but also as a co-factor that binds to an alternate location on the enzyme. The biological fixation of CO_2 upon complexation with Mg^{2+} at the active site of the enzyme and the involvement of carboxyarabinitol-bisphosphate carboxylase confirmed the CO_2 intervened carbamoylation of the amino group of the lysine residue during this process [8].

Carbamate Toxicity

Carbamate leads to cholinergic as well as non-cholinergic toxicity in humans. Exposure to carbamates in humans inhibits the release of acetylcholinesterase and targets human melatonin receptors to a degree lesser than the other insecticides [9]. In 1985, aldicarb found in watermelon samples was found to cause an outbreak of pesticide food poisoning in California that sickened over 2,000 individuals and resulted in a temporary ban on the sale of watermelons [10]. Owing to their toxicity, they are considered in the USEPA's 1992 priority list for limiting their use [11]. Carbamates like carbaryl, aldicarb, carbofuran, methiocarb, and propoxur have been found to affect the reproductive health of an individual and exhibit mutagenic and teratogenic properties in cases of long-term exposures [12].

Carbamates and their metabolites have been reported to disrupt the activity of steroidal hormones and alter their normal functioning. They have also been found to mimic hormones, binding their receptor sites and disrupting the expression of the gene. Carbamates behave as agonists of the human estrogen and progesterone receptors [13]. The release of reactive oxygen species (ROS) during carbamate metabolism leads to oxidative stress. These ROS have been shown to cause cytotoxicity when they interact with lipids, proteins, or DNA [14]. Carbamates and their metabolites are carcinogenic too, prompting chromosomal aberrations [15]. The clinical consequence ultimately depends upon the dosage, kind of carbamate, and the route followed by the carbamate inside the living system. Acute carbamate poisoning effect is observed if it is inhaled directly. One starts recovering from mild carbamate toxicity without any severe damage after 4 hours of exposure.

Carbamate Toxicity Remediation Modes

Pertaining to the toxicity highlighted above, the removal of carbamates from water bodies becomes a matter of serious concern and raises a contextual necessity to develop the detoxification procedures. The degradation of carbamates at the primary level was performed by Lemley et al. in 1984 in the presence of aqueous acid [16]. The authors described the hydrolytic cleavage of carbamates on the ion-exchange resin surface. In another strategy proposed by Union Carbide, these deposits from water supplies were eliminated via adsorption on neutral carbon [17]. Since then, a variety of methods have been explored for their potential to either degrade or eliminate the carbamate pesticides from water, soil, food, and air. All the employed methods along with their advantages and disadvantages are clearly illustrated in the text below.

Adsorption

Adsorption is considered one of the easiest, cheapest, and most efficient modes for pesticide removal from wastewater. A large array of adsorbents are still used to remove metal ions and wastes from water due to their high reuse potential, low investment cost, simple operating designs, and regeneration properties. Pesticides can be eliminated by using various adsorbents including marine sediments, hyper-crosslinked polymer (HXLGp), and powdered activated carbon.

Activated Carbon and Charcoal

The widely used adsorbent for pesticide removal was activated carbon or activated charcoal, because of its high adsorption capacity towards organic pollutants. In 2005, Yang et al. studied the adsorption of methomyl pesticides on marine sediments [18]. The adsorption equilibria were achieved in 10h at 25°C and the adsorption data fitted well into Freundlich adsorption isotherm. El-Geundi et al. too examined the adsorption capacity of cotton-stalk-derived carbon towards methomyl solutions [19]. According to the study, methomyl elimination proceeded rapidly initially and then declined as contact duration increased. For the highest concentration, total adsorption reached a maximum of 87.9%. However, the widespread commercial use of activated charcoal is somewhat restricted because of the relatively high cost and difficulty in the regeneration process. Thus, the use of waste materials as low-cost adsorbents has become an attractive approach towards carbamates remediation.

Khuram Shahzad et al. used *Prunus dulcis* shells as precursors for activated carbon synthesis since they are a sustainable and non-toxic material [20]. His research looked into the adsorption-desorption process of the benzimidazole-based fungicide methyl benzimidazole carbamate (MBC), as well as its removal using almond shells. The remediation assay was shown to be concentration-dependent. Complete elimination of MBC was observed with a maximum concentration of 2.5 ppm.

Biosorbents

Biochar is a rich source of carbon, made from the pyrolysis of biomass sources in an oxygen-limited environment. Biochar treatment has been found to be effective at reducing a wide range of organic and inorganic pollutants found in the soil in their most mobile forms [21]. A study by Yu et al. found that incorporating biochar into the soil can significantly reduce pesticide bio-availability, and can be used to limit the carbamate pesticide uptake by plants from polluted soils [22]. Peng Zhang et al. in 2013 elucidated the influence of biochars with varying ash content onto the fate of two pesticides carbaryl and atrazine [23]. The adsorption and hydrolytic cleavage studies have been performed on ashed as well as de-ashed biochars and results obtained with de-ashed biochar were found to be better than the ashed one. Biochars promoted the chemical degradation of pesticides as well as their bio-degradation.

Another bio-sorbent orange-peel has been used for adsorptive removal of carbofuran by Chen et al. [24]. The adsorption equilibrium was established within a short span of 1 hour. Adsorption results fitted well into the Langmuir adsorption model and the maximum adsorption capacity achieved was 84.49 mg/g of adsorbent. Another example of biosorption includes usage of *Pistia stratiotes* biomass for adsorptive removal of the carbaryl by Soumya Chattoraj et al. [25], lignocellulose [26] for adsorption of isoproturon, dimetomorph carbamates, and orange bagasse to remove methomyl by Thiago Ruiz Zimmer [27]. The residue obtained upon the extraction of orange juice is termed orange bagasse, which quantizes approximately 50% of the mass content of the fruit, thus serving the dual purpose of waste disposal and pesticide remediation. The presence of hydroxyl, carbonyl, and amino functional groups in the orange bagasse, makes it more accessible to pesticides due to chemical interactions [28]. All the examples cited above showed that the biosorbents are viable options for water pollution remediation and to get rid of hazardous carbamates from water bodies.

Nanomaterials

Natural adsorbents are counted as potential adsorption materials because of their low-cost and eco-friendly nature [29]. An ideal absorbent should possess physical and chemical stability in addition to the porosity and a higher surface area to volume ratio. Nano-sized particles of several metal-oxides have been found to fulfil all these requirements and thus have been established as efficient adsorbents towards pesticide removal from wastewater.

Carbon nanotubes (CNTs) discovered in 1991, are counted as novel nano-adsorbents. This category of carbon-based adsorbents includes single and multi-walled carbon nanotubes. The physiochemical and electrical properties of these materials are greatly influenced by their unique peculiar structure. These features were better known for providing applications in various analytical fields. Atrache et al. in 2015, has implied solid-phase extraction in combination with liquid chromatography for the detection of carbamates in different surface water samples [30]. Because of its strong interaction with organic molecules via electrostatic, hydrogen bonding, hydrophobic, and van der Waals interactions, carbon nanotubes (CNTs) have emerged as a promising candidate for adsorption-cum-removal of carbamates. These interactions reside because of their characteristic hollow and layered nanosized structure. The researcher also discovered that the aromatic carbamates (those with benzene rings) were significantly adsorbed on CNTs due to pi-pi bonding interactions. The key benefit of this study was that it did not require any sophisticated sample pre-treatment to extract the *N*-methylcarbamate insecticides.

In 2015 A. Derbalah et al. synthesized alumino-silica based 3-D mesocage particles having a cavity of 7.0 nm [31]. This large-sized cavity and surface area in addition to several active acidic sites proved to adsorb a better amount of carbamate pesticides in its pockets from contaminated water. The guest accommodation property of the mesocage collector cavities established these nano pocket collectors as an effective and recyclable adsorbent for the carbamates. However, the reversibility of adsorption was not found satisfactory upon their reuse.

Due to cost effects and difficulty in preparing high-quality fine oxide powders, these materials bring about new challenges in the field of environmental remediation.

Disinfection

In classical times, chlorination, ozonation, and UV radiations were employed to degrade or destroy the harmful pesticidal properties of carbamates in the aquatic environment. Chlorination with Cl_2, ClO_2, and chloramine are primary reagents among these, which have been worked upon for disinfection of pesticide-contaminated water.

Chlorination and Ozonation

A straightforward methodology for the remediation of the carbamates from drinking water by disinfection using Cl_2 ClO_2, and O_3, was introduced by Mason et al. in 1990 [32]. Mason and his co-workers performed the degradation of some popular carbamates like aldicarb, methomyl, carbaryl, and propoxur using these disinfectants. It was found that two aromatic carbamates carbaryl and propoxur remain unaltered to chlorination, and all of the carbamates were neutral to the ClO_2 treatment, even when taken in excess. However, all the carbamates had shown better reactivity with ozone under similar reaction conditions. Since ozone is a strong oxidizing agent which undergoes self-decomposition in aqueous media to release hydroxyl radicals, ozonation had been found as an effective technique for the degradation of the carbamates.

Methiocarb (3,5-dimethyl-4-(methylthio)phenylmethyl carbamate) also called mesurol, is classified as the most hazardous effluent by World Health Organization. A report in 1999 published by the "Joint meeting of pesticides residues" stated that methiocarb gets degraded into other derivatives in an aqueous environment, thus leading to secondary level toxicity [33]. F. Tian et al. had reported the kinetics and pathways for the reaction of methiocarb with free chlorine and chlorine dioxide [34]. The rate constant for the degradation reaction of methiocarb with chlorine was found to be $(2.42 \pm 0.09) \times 10^8 M^{-1}s^{-1}$. The major concern reported by authors was that on chlorination with either disinfectant (Cl_2, ClO_2, NH_2Cl), methiocarb oxidized to other toxic degradation products under different pH conditions. In any case, these strategies were not able to eliminate the carbamates pesticides completely from the environment because of the stability of by-products.

Zhimin Qiang (2013) explored the use of monochloramine towards the remediation of methiocarb, under simulated water treatment conditions [35]. The reactivity shown by monochloramine with methiocarb pesticide was of the first order in kinetics. But the order of the reaction was not found to be the same for all the carbamates under different conditions of pH. Methiocarb followed second-order reaction kinetics, during its disinfection with free chlorine and ClO_2. Monochloramine (NH_2Cl) has been progressively utilized as an alternative to free chlorine to show disinfectant qualities. But it too leads to the minimal formation of the by-products itself, thus falls under the category of secondary disinfectants [36].

A. Cruz-Alcalde et al. in 2017, evaluated the degradation of methiocarb and all its transformation products [37] by ozonation. In the degradation experiments, the methiocarb could be eliminated with direct ozone treatment. But the degradation intermediates left behind after ozonation could be oxidized only in the presence of the hydroxyl radicals. A possible explanation for this is that the presence of $OH^{\cdot}$ radicals could significantly enhance the indirect degradation routes of the methiocarb carbamate.

To conclude, all these disinfection techniques are not able to degrade the whole of the carbamate and its metabolites, thus new strategies are now being adopted for the remediation of the carbamates from wastewater.

Ultraviolet (UV) Rays in Conjunction with Chlorination

UV radiations have also been reported to be involved in the photooxidative degradation of pesticides. The frequency ranges from 10 to 400 nm for UV and from 100 to 200 nm in the case of vacuum ultraviolet rays (VUV). The use of this technique is favourable as not only the toxic pollutants but the water too get oxidized under the influence of energetic VUV rays generating the reactive oxygen species to disrupt the molecules [38]. The significance of these two techniques impelled Yang et al. to combine them to bring a synergistic effect into the degradation process [39]. Thus, this technique is now regarded as an advanced method of oxidative remediation. The outcomes reveal that this technique can degrade all the pesticides rapidly to a remarkable extent. In a degradation study by Fang et al. (2014), the ·OH and ·Cl radicals produced upon ozonation strongly aided the degradation of benzoic acid, carried out under the combined influence of UV-rays and chlorine [40]. As per the study by Dong et al. (2017), the rate of degradation of chloramphenicol by UV-chlorine combination was 10 times faster than sole UV and 2 times faster than the sole chlorination process [41].

Redox Methods

Reduction by Zerovalent Iron Powder (ZVIP)

By the late 1980s, ZVIP was considered one of the most efficient reducing agents for the degradation of organic compounds [42]. Many chemicals may easily oxidise ZVIP (Fe^0) to Fe^{2+} ions. There exists a Fe^0 - Fe^{2+} redox pair with the standard potential of 0.44 V that leads to metal corrosion in humid conditions, and this makes Fe(0) a reducing agent. Matheson and Tratnyek (1994) were the first to reduce alkyl halides (RX) with zerovalent iron [43]. The reductive dehalogenation was carried out in the aqueous solution as follow:

$$RX + 2e^- + H^+ \rightarrow RH + X^-$$
$$Fe^0 + RX + H^+ \rightarrow Fe^{2+} + RH + X^-$$

Later, A. Ghauch et al. did a study to see how effective ZVIP therapy is for carbaryl degradation under aqueous conditions [44]. The authors found that this technique could be used effectively with the carbaryl highest concentration (40 mg/L), at room temperature, and neutral pH. The rate of carbaryl degradation increased with the increasing amount of ZVIP. Thus, the low-cost and less-toxic ZVIP treatment has proven to be an effective method for pesticide removal from ground and surface water.

Fenton Process

Fenton et al. in 1894 used a mixture of ferrous salts and hydrogen peroxide for the oxidation of tartaric acid to form dihydroxy maleic acid [45]. This mixture was in turn, termed as a Fenton reagent and the process as Fenton process. Later in 1934, Haber and Weiss had proposed the formation of hydroxyl radicals from the iron-catalyzed decomposition of hydrogen peroxide, which can degrade a variety of simpler and complex molecules [46]. It was found to be a quite effective technique, reflected by a history of applications in the removal of contaminants from the wastewater [47]. Watts and colleagues were the first to use Fenton and Fenton-like procedures to degrade the common organic pollutants like dieldrin, pentachlorophenol, trifluralin, hexadecane, and trichloroethylene [48]. Apart from ferrous ions (Fe^{2+}), other metal ions such as Cu^{+}, Ti^{2+}, Cr^{3+}, Co^{2+}, have also been found to show oxidative properties near to that of Fenton's reagent [49]. Neyens and Baeyens (2003) explained the complex chain reactions which were involved in the generation of hydroxyl radicals and their subsequent reactivity [50]. Fenton process does not require any external energy input to activate H_2O_2 and have relatively high-efficiency values and shorter degradation time interval. Thus, the classical Fenton treatment emerged as a cost-effective method of waste remediation. Depending on the mechanism involved in the generation of hydroxyl radicals, the Fenton process can further be classified into subgroups as per Table 1.

The main principle behind the classical Fenton treatment involves the reaction of ferrous ions with hydrogen peroxide for the in-situ production of the hydroxyl radicals as shown in equation (1) below.

$$Fe^{2+} + H_2O_2 + H^+ \rightarrow Fe^{3+} + \cdot OH + H_2O \quad (1)$$

$$OH + Fe^{2+} \rightarrow OH^- + Fe^{3+} \quad (2)$$

These radicals can act as a very strong oxidant to degrade the organic pollutants. The Fe^{3+} ion reacts with H_2O_2 to give Fe^{2+}-hydroperoxyl ion complex as below:

Table 1. A comparative account of Fenton processes towards wastewater remediation

S. No.	Fenton Techniques	Source and Conditions	Merits & De-merits
1.	Homogeneous Fenton Process	H_2O_2+Fe^{2+} or H_2O_2+O_3 The homogenous mixture consists of metals such as Fe^{3+}, Cu^{2+}, Mn^{2+}, Co^{2+}, and Ag^{+} along with ligands (citrate, oxalate, EDTA, humic acid, and ethylenediamine succinic acid).	Difficulty in the formation of ferric hydroxide sludge at high pH values, low catalyst recyclability, and reusability, high energy consumption, and operative at pH levels below 4 only.

Table 1. (Continued)

S. No.	Fenton Techniques	Source and Conditions	Merits & De-merits
2.	Photo-Fenton Process	H_2O_2+UV, O_3+UV, H_2O_2+Fe^{2+}+UV, or TiO_2+UV	It involves the use of high-energy UV radiations for a long time which increases the cost and risk factor.
3.	Sonochemical Fenton Process	H_2O_2+Fe^{2+} or H_2O_2+O_3 with Ultrasound radiation	This method, too, has limitations such as excessive energy usage and greater costs.
4.	Electrochemical Fenton Process	Electrogenerated H_2O_2 from electric current	The efficiency of the current generation gets reduced under highly acidic conditions (pH < 3). Also, the rate of formation of H_2O_2 is slow in the process due to the poor solubility of oxygen in the water.
5.	Anodic Fenton Process	Electrogenerated H_2O_2 from electric current with the use of an ion-exchange membrane or salt bridge (NaCl).	The main advantage of this technique is that it enables the reaction to occur even at lower pH values (acidic, pH<3). The efficiency of the process is limited to various factors such as the nature of the electrode, pH range, the concentration of the catalyst and electrolytes, dissolved oxygen level, current density, and the temperature of the solution.
6.	Fenton with iron-containing compounds (Heterogeneous Process)	Zero valent iron, Fly ash, iron sludge, iron pillared clays (kaoline and sepiolite), iron minerals like magnetite and goethite.	Low cost and high surface area with rich active sites of materials and post easy separation are a few merits of the process.

$$Fe^{3+} + H_2O_2 \rightarrow Fe\text{-}OOH^{2+} + H^+ \quad (3)$$

The decomposition of $Fe\text{-}OOH^{2+}$ led to the generation of ferrous ions and hydroperoxyl ions.

$$Fe\text{-}OOH^{2+} \rightarrow HO_2\cdot + Fe^{2+} \quad (4)$$

Equation (3) and (4) are characteristic of "Fenton-like" reactions. The ferric or ferrous ions may react with hydroperoxyl radicals as:

$$Fe^{2+} + HO_2\cdot \rightarrow Fe^{3+} + HO_2^- \quad (5)$$

$$Fe^{3+} + HO_2\cdot \rightarrow Fe^{2+} + O_2 + H^+ \quad (6)$$

The classical Fenton treatment was found to be simple to use, although it had a few flaws. The Fenton treatment process was rendered unmanageable due to the use of ferrous salts, which were very hygroscopic and easily oxidizable. Some new approaches were introduced to make it more reliable. The electrochemical Fenton method, photochemical Fenton method, and anodic Fenton method served as the appropriate examples in this direction, as they increased the efficiency of the classical Fenton treatment in one or another way.

Photo-Fenton Process

Irradiation with UV-Vis rays during the Fenton process, enhanced the rate of degradation of organic pollutants. In this modification, the photolysis of Fe^{3+}-hydroxyl complex was carried out to generate $^{\cdot}OH$ radicals, which is known as the photo-assisted Fenton process,

$$Fe\,(OH)^{2+} + h\nu\,(\lambda < 400\text{ nm}) \rightarrow Fe^{2+} + {}^{\cdot}OH$$

It undergoes a metal charge transfer excitation process. In addition, direct photolysis of H_2O_2 leads to the generation of hydroxyl free radicals that can oxidize the organic pollutants readily.

$$H_2O_2 + h\nu \rightarrow 2\,{}^{\cdot}OH$$

UV light and sunlight are commonly employed light sources in the photo-assisted Fenton process. In 2008, Poulopoulos et al. carried out the degradation of 2-chlorophenol in UV/H_2O_2 as a photo-Fenton system [51]. The oxidation capacity of the Fenton process was found to increase in the presence of UV light, and around 95% of the total organic carbon (TOC) was removed within a short span of 150 minutes. The successful usage of sunlight as an alternative to UV light in the photo-Fenton process has been demonstrated by Kuo et al. in 2010 [52] and Liu et al. in 2013 [53]. Pliego et al. (2014) carried out a comparative study of the classical and photo Fenton processes towards the remediation of the antidepressant sertraline under aqueous conditions [54]. The removal efficiency was found to be 2.5 folds of that in the classical Fenton, indicating the increased oxidative tendency of the solar-Fenton process over the conventional one. However, the use of UV irradiation equipment in the UV-Fenton process is filled with issues like higher energy consumption, health risks related to UV exposure, and the cost [55].

Huston et al. in 1998, reported the degradation of active pesticide ingredients like aldicarb, alachlor, methyl-azinphos, atrazine, carbofuran, captan, and dicamba under photo-Fenton conditions [56]. Degradation of all the ingredients and complete mineralization thereafter was observed in a short interval of 30 mins and 120 mins respectively. The authors also discovered the inhibitory role played by the presence of adjuvants (inert ingredients) in the degradation of the active ingredients of commercially available carbamate products under the photo-Fenton method. The presence of adjuvants had little or no effect on the rate of degradation of carbofuran and a paramount effect on the degradation of alachlor. Moreover, the degradation rate of the pesticide active ingredients was affected considerably by the pH of the solution and the degradation followed the first-order kinetics.

Electrochemical Fenton Method

The electro-Fenton process has been accounted as one of the productive oxidative-degradation techniques that rely on the electrolytic generation of H_2O_2 at the cathode continuously. The addition of an iron catalyst results in the reduction of oxygen on graphite cathode, as well as alternative cathodes like mercury pool and carbon-felt to produce the oxidant [57].

The reduction of oxygen on graphite in the acidic medium can be represented by the following electrochemical reactions:

$$H_2O \rightarrow \cdot OH_{ads} + H^+ + e^-$$
$$O_2(g) + 2H^+ + 2e^- \rightarrow H_2O_2$$
$$Fe(s) \rightarrow Fe^{2+} + 2e^-$$
$$2H_2O + 2e^- \rightarrow H_2 + 2OH^-$$

The key advantage of this electro-Fenton procedure over the classical Fenton process was the in-situ generation of H_2O_2. It avoids the risks pertaining to the transport of H_2O_2 and its storage [58].

Mutlu Sonmez Celebi et al. (2015) employed the electro-Fenton approach to decompose carbaryl utilizing Pt or boron-doped diamond anodes and carbon felt as the cathode [59]. Mineralization of the pesticide solutions was monitored in terms of the total organic carbon (TOC) content analysis. It was discovered that under optimal operational conditions, about 90% of TOC could be removed in just 2 hours. The carbaryl degradation reaction followed pseudo-first-order kinetics. Furthermore, the degraded products like aromatic intermediates, short-chain carboxylic acids, and inorganic ions were quantified by the means of HPLC and GC-MS techniques.

In 2016 Selva et al. carried out the electrochemical oxidation of the carbamate pesticide propoxur, utilizing a boron-doped diamond electrode [60]. He investigated the irreversible oxidation of propoxur over the pH range 2 to 7, in aqueous solutions. The oxidation process was carried through a single electron transfer without the interference of any proton. The adopted method is appropriate and well suited for the detection and quantification of propoxur in the micromolar range using differential pulse voltammetry. Moreover, the proposed sensor being cost-effective and reproducible is well suited for routine laboratory and field investigations.

Anodic Fenton Method

This modified Fenton method was developed in the laboratory by Saltmiras and Lemley in 2001 [61]. It outperformed the classic and electro-Fenton processes in terms of degradation efficiency. This methodology has been successful in the degradation and detoxification of toxic chemicals ranging from thiourea, 2,4-DPAA, to carbamates like carbofuran and carbaryl.

In the anodic Fenton methodology, the anodic half-cell generates ferrous ions by the oxidation of iron, and reduction of water takes place in the cathodic half-cell. H_2O_2 was injected into the anodic half-cell continuously. The H_2O_2 helped in the generation of hydroxyl radicals when it oxidizes the ferrous ions to ferric ions.

At anode: $Fe \rightarrow Fe^{2+} + 2e^-$

$Fe^{2+} + H_2O_2 \rightarrow Fe^{3+} + {}^-OH + {}^{\cdot}OH$

At cathode: $2H_2O + 2e^- \rightarrow H_2 + 2OH^-$

To widen the boundaries and to enhance the practical utility of anodic Fenton treatment (AFT), a major change was introduced by Wang et al. in 2003 [62]. The use of a salt bridge was replaced by an anion exchange membrane in the AFT model. The modified AFT method not only increased the degradation percentage of the carbaryl but also reduced the COD value of the sample solution. The degradation of the carbaryl was assured by the detection of degradation products in the GC-MS of the sample solution. On a similar note, the authors also reported the oxidative degradation of carbofuran using membrane AFT [63]. The kinetics of carbofuran degradation at all temperatures was found to follow the AFT model. The degradation rate of carbofuran in the solution improved with an increase in initial concentration and treatment temperature. As per GC-MS analysis, four degradation products were produced upon anodic Fenton treatment of the carbofuran. The results showed that modified AFT was effective in improving carbamate biodegradability.

In 2003, Knepper et al. has posed a problem over the presence of adjuvants like surfactants and organic solvents (used in the formulation of pesticides) in the wastewater [64]. The presence of surfactants in the pesticide-contaminated wastewater has been reported to influence the photosensitization and degradation process by micelle solubilization [65]. An inhibition in photooxidative degradation of carbaryl pesticide by surfactants like sodium dodecyl sulfate, hexadecyltrimethylammonium bromide, and poly-oxyethylene-dodecyl ether was studied by Bianco Prevot et al. [66]. Barrios et al. [67] too reported the reduction in the rate of photo-assisted degradation of naphthalene in the presence of Triton X-100 surfactant. These surfactants inhabiting the pesticide accumulated wastewater can significantly affect the efficiency of pesticide degradation carried out by any mode. The surfactants with anionic character have been discovered to either enhance or suppress the pesticide photodegradation phenomenon [68]. Taking this into account, Kong and Lemley (2006) addressed the effect of surfactants on the degradation routes for carbaryl undergoing AFT [69]. Based on the control experiments, Triton X-100, a non-ionic surfactant, decreased the degradation rate of carbaryl significantly upon increasing its concentration from 20 to 1000 mgL^{-1}. The authors formulated that the surfactant might have led to the depletion of $\cdot OH$ radicals during the complex formation between the surfactant, Fe^{3+} ions, and the carbaryl. And this deficiency of $\cdot OH$ radicals, in turn, minimizes the hydroxy radical-directed degradation of carbaryl moiety.

Lemley et al. (2008) also studied the adsorption and degradation behaviour of carbaryl, paraquat, and mecoprop pesticides on montmorillonite clay [70]. They found that the tightly held pesticide was degraded to a lesser extent during the anodic-Fenton process, due to lesser exposure to hydroxyl radical and those which are loosely bound could be degraded easily.

The presence of natural organic matter (NOM) has a considerable impact on Fenton-related processes towards the removal efficiency of the contaminants. The NOM content in waste-water bodies has been shown to limit the decomposition of contaminants in many studies [71]. On contrary, an enhancement in the degradation rate of organic compounds such as phenols and acids has also been observed in the presence of NOM [72]. But it was reported to be controversial in the case of carbamate degradation studies by Peng Ye and Ann T. Lemley [73]. A kinetic model for the degradation of carbaryl pesticide using AFT was studied in the artificial soil system to resolve this issue. The artificial soil that was used as a mimic of natural soil consisted of silica, kaolinite clay, with varying degrees of humic acid, and widely used carbaryl insecticide. The results revealed that the more the soil was enriched with OM, the greater was the inhibition of the pesticide degradation under AFT conditions. The humic acid content in the soil was correlated to pesticide degradation. It slowed down the pesticide degradation phenomenon probably due to its higher adsorption capacity. The study revealed that a portion of the pesticide gets adhered to the solid surface and is not susceptible to degradation.

The role of NOM in the Fenton degradation system was further explored by Fan et al. (2011) [74]. They conducted the degradation of methomyl in the acidic environment under NOM-influenced Fenton reagents. The degradation was found to be efficient up to 20mg/L, 2mM, and 2mM concentrations of methomyl, ferrous ions, and H2O2 respectively when carried out in the absence of humic acid. A further increase in the concentration of ferrous ions didn't bring any significant change. However, in the presence of humic acid, a competition arises between the humic acid and methomyl, for hydroxyl radicals. But the results improved with a higher dosage of ferrous ions, as the humic acid gets oxidized and cleaved to produce more and more hydroxyl radicals which in turn degraded the methomyl.

Photo-Remediation

Heterogeneous Photocatalytic Ozonation

Photocatalytic ozonation is another potent method employed towards oxidative degradation of the toxins inhabited in wastewater [75]. When several photocatalysts were combined with ozone, a synergistic effect was observed. Commonly used catalysts in photocatalytic ozonation comprise oxides of transition elements (like Ti, Mn, Fe, Ce), transition metals (Cu, Ru, Pt, and Co) immobilized onto these solid oxides, or zeolites, and carbon [76]. The significance of this conjunctive-oxidation technique lies in that it can decompose even the poorly biodegradable organic molecules. Thus, it improved the potential for the biological degradability of pesticides.

M. J. Farre et al. carried out Fenton-ozone treatment of the pesticides like alachlor, atrazine, isoproturon, diuron, and pentachlorophenol [77]. Partial degradation of the pesticides generated intermediates that could be degraded readily by biological means. Thus, a dual approach has been adopted for pesticide remediation.

Rajeswari et al. studied the photocatalytic ozonation of carbaryl carbamate in the presence of titanium dioxide [78]. The TOC content and COD values declined simultaneously indicating the decay of the carbamate under photocatalytic ozonation. A total of 92% reduction in COD value and 76.5% reduction in TOC content were achieved in 3 hours. However, the synergistic effect was lost under basic conditions due to the self-decomposition of ozone. The primary mechanism in oxidation processes involves the generation of hydroxyl free radicals. Theoretically, the photocatalysts possess enormous potential to transform photon energy to chemical energy for the degradation of complex organic molecules, and thus the concurrent use of photocatalysis and ozonation synergistically improves the formation of free radicals and thereby speed up the degradation phenomenon. The authors also reported the degradation of another carbamate carbendazim under photocatalytic ozone treatment [79].

Heterogeneous Photocatalytic Degradation

Owing to its non-toxic nature, heterogeneous photo-catalysis is counted as an efficient and excellent procedure for pesticide elimination from wastewater. Purification of water using TiO_2 as a photocatalyst dates back to the late 1990s [80]. This method gained much importance as it includes minimal expense and plates can be reused without any loss in activity. But its separation after use was a major shortcoming for the powdered form of TiO_2. The efficiency of this treatment depends upon factors like the illumination of the source, the thickness of TiO_2 film, the surface area of the photocatalyst, and the concentration of substrate used. K. Tennakone et al. attempted the mineralization of the carbofuran carbamate and found that the complete mineralization could be achieved only after 15h of UV irradiation under the optimized experiments [81]. The pre-requisite of a high-energy UV light source for water decontamination remained the only disadvantage of this technique.

To reduce the risk hazards associated with the long-term exposure of UV radiations during photocatalysis, a wide range of photocatalysts were employed to achieve photodegradation with higher wavelength UV-Visible rays. The photocatalysts that have been typically employed in the photodegradation of alachlor, aldicarb, methomyl, carbaryl, primicarb, and metolachlor carbamate pesticides include transition metal oxides and sulfides such as TiO_2, ZnO, Fe_2O_3, MnO_2, CdS, and ZnS, Fe(III) aqua complexes, ZnO-$Na_2S_2O_8$, and TiO_2 nanoparticles doped with $CdSO_4$.

Among them, TiO_2 and ZnO have been frequently used [82-83]. ZnO appeared to be more efficient than TiO_2, only if used in high concentrations (greater than 0.2 gL^{-1}). Moreover, ZnO absorbs at a larger wavelength of 425-440 nm. But the band gaps of both TiO_2 and ZnO photocatalyst are nearly identical being 3.1 and 3.2 eV respectively. Furthermore, the photocatalytic degradation with ZnO was more productive while carried out under acidic conditions, as there are chances of photodecomposition and photo corrosion of ZnO in a basic medium [84]. In a comparative study conducted by Jose Fenoll et al., the residual levels of the carbofuran insecticide, after 240 minutes of treatment with ZnO, degussa, rutile, and anatase forms of TiO_2, were found to be 0.1, 22.4, 62.8, and 68.4 μgL^{-1} respectively [85]. Pseudo-second order kinetics was followed in each case and oxidative-degradation was established as the primary cause of pesticidal remediation.

Numerous examples of TiO_2 and ZnO-directed carbamate pesticide degradation are cited in the literature. L. Lhomme et al. in 2008, reported the degradation of the cyproconazole and chlortoluron pesticides solution in pure water with heterogeneous TiO_2 photocatalyst [86]. The photocatalyst was able to mineralize the pesticide chlortoluron completely and cyproconazole partially. In another work, Barakat et al., introduced $CdSO_4$ nanoparticles while carrying out the heterogeneous degradation with TiO_2 and an overall enhancement in the photocatalytic activity was achieved [87]. The nanoparticles were able to eliminate the methomyl carbamate successfully, up to an effective concentration of 2,000 mg/L, in less than an hour.

In 2011, decomposition of the carbaryl pesticide was studied in the presence of $\cdot NO_3$ radicals and the results were monitored with the help of VUV equipped with a mass spectrometer [88]. The reaction of nitrate radical with carbaryl followed a chemical route similar to that followed by polycyclic aromatic hydrocarbons in the upper atmosphere. Thus, this pathway, which is analogous to the one which already persists in nature, can be counted as a green carbamate decay method.

In 1998, Prevot et al. investigated the effects of surfactants onto the photo-degradation of carbaryl in the presence of anatase form of TiO_2 particles [66]. Above the critical micellar concentration, the surfactants competed with the substrate for the active sites of the catalyst and the surfactants thus were found to have an inhibitory influence on the rate of degradation. But with longer durations, the surfactants started degrading, freeing the adsorbed surface and re-allowing the substrate molecules to access the catalytic surface again, and the rapid decay of the carbaryl was observed. Another report of the inhibitory action of micelle formation onto the hydrolytic cleavage of carbofuran has been studied by Manuel et al. [89]. The ionic and non-ionic surfactants were found to affect the decomposition of carbofuran differently. Cationic surfactants were found to enhance the decomposition rate, whereas the anionic and non-ionic surfactants served as inhibitors for the same. The catalysis and inhibition effects shown by surfactants were attributed to the association of the carbofuran and micellar core and the exclusion of hydroxy anion from the Stern layer.

A report by Andelka Tomasevic et al. in 2017, compared the degradation of carbofuran and its commercially available products in the presence of ZnO catalyst [90]. A comparison of the effects of inert substances on the photodegradation of furadan 35-ST was also carried out. The degradation rate was found to be higher for pure carbofuran than that of the commercial samples, indicating the interference caused in the degradation due to barrier formation by the associated ingredients.

Photo-Fenton

Described as a sub-heading under Fenton process.

Photosensitized Degradation

Photo-sensitized degradation involves the addition of a photosensitizer that can absorb light and then transfer it to the photocatalyst which will be able to carry out the degradation process by a series of reactions. The dyes serve the aforementioned purpose well. The photosensitization has also been found to increase the overall quantum yield of the process.

W. S. Kuo et al. carried out the photocatalytic degradation of carbaryl in the presence of methylene blue (MB) dye as a sensitizer [91]. It was observed that the addition of the MB dye, corresponding to 1% of the carbaryl concentration, increased the mineralization efficiency to 26.2 and carbaryl removal to 66.2 percent. The authors accounted hydrolysis, hydroxylation, and quinone formation, as the principal steps involved in the carbaryl degradation in the presence of sunlight and MB dye. The authors have also reported the use of MB and Rose Bengal (RB) dyes as sensitizers in the photocatalytic degradation of carbofuran. Under similar conditions, the RB dye had a stronger influence on the breakdown of carbofuran. With the addition of 2 mol% RB dye, 70 percent of the pesticide was degraded, while only 28 percent of the pesticide was mineralized. Another report of the degradation of methiocarb and ethiofencarb carbamates photosensitized by anthraquinone in an H_2O-CH_3CN solvent mixture has been studied, by A. Galadi et al. [92]. Complete degradation of the carbamates was observed upon irradiation for 0.5-0.8 hours.

Bioremediation

The ability of microorganisms to degrade the complex chemical structure into a simpler form is referred to as bioremediation. Certain microorganisms like bacteria have evolved specific degradation pathways to counteract the unavoidable presence of toxic chemicals. According to another hypothesis, a single bacterium is unable to degrade the carbamate or mixture of carbamates present in the given soil or water sample, thus a bacterial consortium may be held responsible for the whole degradation process [93]. In a few cases, metabolic synergism has been found to operate, which means the metabolites of one microbe can be acted upon by another microbe and mineralized [94]. As a case, in Propoxur degradation reported by Kim et al., Pseudaminobacter strain hydrolyzed propoxur to iso-propoxyphenol, which is further acted upon by Nocardioides to give the ultimate degradation product.

Bacteria, fungi, and algae are among the microbes employed towards carbamate remediation from wastewater [95-96]. Species from bacterial strains Aminobacter, Arthrobacter, Bacillus, Burkholderia, Cupriavidus, Enterobacter, Escherichia, Micrococcus, Novosphingobium, Paracoccus, Pseudomonas, Rhizobium, Rhodococcus, Sphingbium, Sphingomonas, Stenotrophomonas and fungal strains Acremonium, Mucor, Trametes, Trichoderma have been explored for their potential in carbamate pesticide degradation [95].

Carbamate degradation by the microbes leads to their partial or complete mineralization (Table 2); wherein the first step involves the hydrolytic cleavage of ester or amide linkage catalyzed by carboxylesterase enzyme and the second step involves the degradation of the remaining hydrocarbon chain by C1 metabolism or another oxidation mode [97]. As a case, hydrolysis of methomyl produces carbamic acid and alcohol, which is degraded further into methylamine and CO_2, which can be used as a carbon or nitrogen source by the microbial cells. Besides this, certain other factors like soil characteristics, pH, temperature, and nutrient content, also play a major role in the degradation phenomenon.

Table 2. Biodegradation of carbamate pesticides by various microbes [95-96, 98]

Microbe Category	Microbe Name	Carbamate degraded	% degradation and mineralization
Bacteria	Arthrobacter sp.	Aldicarb, Carbaryl, Propoxur,	100%, 80%, 37%
	Aminobacter sp.	Methomyl, Oxamyl	88%, 99%
	Bacillus sp.	Methomyl	88.25%
	Burkholderia sp.	Carbofuran, Carbaryl	90%, 99%
	Cunninghamella elegans	Aldicarb	66%
	Cupriavidus sp.	Carbofuran	98%
	Enterobacter sp.	Aldicarb, Carbofuran, Carbaryl	100% for all with immobilized cells and 75% with free cells
	Neisseria	Propoxur	60%
	Novosphingobium sp	Carbofuran	95%
	Paracoccus sp.	Carbofuran	88%
	Pseudomonas sp.	Aldicarb, Carbaryl, Methomyl, Oxamyl, Propoxur	100%
	Sphingbium sp.	Carbofuran	99%
	Sphingomonas sp.	Carbofuran	95%
	Trichoderma	Aldicarb, Oxamyl	89%
Fungi	Acremonium	Carbofuran, Carbaryl	69.7%, 75.44%
	Mucor ramannianus	Carbofuran	100%
	Penicillium multicolor	Aldicarb	66%
	Trametes versicolor	Aldicarb, Carbofuran	87%, 99.5%
Algae	C. vulgaris	Oxamyl	69.61
	S. obliquus	Oxamyl	75
	A. oryza	Oxamyl	64.8
	N. muscorum	Oxamyl	54.58

The cleavage of the carbamate's ester or amide linkages by esterases, which is the first step involved in the biodegradation of the carbamates, has been reviewed extensively in the literature [98]. And it has been established on the basis of mechanistic studies that the enzymatic actions play a far important role in the biodegradation of the carbamates [99]. Thus, the direct microbial enzymatic applications in the bioremediation process may prove to be a much more efficient and effective approach. Furthermore, the steps involved in the microbe-derived biodegradation of the carbamate pesticides have been established by carrying genetic manipulations. Genetic manipulations have also helped in increasing the bioremediation potential of the enzymes [93, 95, 98].

Conclusion and Future Prospects

To summarize, the wider use of carbamates in routine agricultural practices necessitates the development of cheap and efficient modes of their remediation from the environment. Much progress has been made in all the remediation fields. For adsorption strategy, a variety of new and nano-range materials have been discovered. The disinfection modes have also been used under integrated modes like UV-Cl_2 and O_3-Cl_2. Several modifications have also been made in the electrochemical and photochemical methods of remediation. Modified Fenton processes like photo-Fenton, sonochemical-Fenton, electrochemical-Fenton, and anodic-Fenton, with increased carbamates degradation capabilities, have surely opened up new remediation possibilities. Doping the classical photocatalysts TiO_2 and ZnO with new conducting materials has served far better results of carbamate degradation. Biosorption and biodegradation, being considered as the cleaner and green approach of remediation, has made substantial progress with the development of metabolic synergism and genetic manipulations, and developed much pace as a mode of carbamate remediation.

As evidenced by the literature data, a single degradation technique is not sufficient enough to eliminate all the carbamates simultaneously and completely, and thus an integrated approach to resolve this issue of carbamate toxicity is the need of the hour. The researchers must address this issue more expeditiously, looking into the processes involved in carbamate remediation, examining the same question from different perspectives, and adopting multiple strategies. A better understanding of the enzymatic actions involved in carbamate degradation pathways, detection and isolation of active genes along with genetic manipulations may help in the development of better degradation techniques in near future.

References

[1] Matsumura, F., 1985. *Toxicology of Insecticides*, Springer, Boston, MA.

[2] Agrawal, A., Sharma B., 2010. Pesticides induced oxidative stress in mammalian systems, *Int. J. Biol. Med. Res.* 1: 90-104.

[3] da Silva, M. K. L., Vanzela, H. C., Defavari, L. M., Cesarino, I., 2018. Determination of carbamate pesticide in food using a biosensor based on reduced graphene oxide and acetylcholinesterase enzyme, *Sens. Actuators B Chem.* 277: 555-561.

[4] Jindra, M., Bittova, L., 2019. The juvenile hormone receptor as a target of juvenoid "insect growth regulators", *Arch. Insect Biochem. Physiol.* 3: e21615.

[5] Vale, A., Lotti, M., 2015. "Organophosphorus and carbamate insecticide poisoning" In *Handbook of Clinical Neurology*, edited by M. Lotti and M. L. Bleecker, Elsevier, Amsterdam, Vol. 131, 149-168.

[6] Meigh, L., 2015. CO_2 carbamylation of proteins as a mechanism in physiology, *Biochem. Soc. Trans.* 43: 460-464.

[7] Shi, D., Caldovic, L., Tuchman, M., 2018. Sources and Fates of Carbamyl Phosphate: A Labile Energy-Rich Molecule with Multiple Facets, *Biology* 7: 34.

[8] Linthwaite, V. L., Janus, J. M., Brown, A. P., Pascua, D. W., Donoghue, A. C. O., Porter, A., Treumann, A., Hodgson, D. R. W., Cann, M. J., 2018. The identification of carbon dioxide mediated protein post-translational modifications, *Nat. Commun.* 9: 3092.

[9] Colovic, M. B., Krstic, D. Z., Pasti, T. D. L., Bondzic, A. M., Vasic, V. M., 2013. Acetylcholinesterase Inhibitors: Pharmacology and Toxicology, *Curr. Neuropharmacol.* 11: 315-335.

[10] Guerrera, A. A., 1981. Chemical contamination of aquifers on Long Island, New York, *J. Am. Water Works Assoc.* 73: 190-199.

[11] U.S. Environmental Protection Agency. *National Survey of Pesticides in Drinking Water Wells*, Phase II Report (EPA570/9-91-020), Office of Drinking Water, Washington, DC, 1992.

[12] Morais, S., Dias E., Pereira M., 2012. "Carbamates: Human Exposure and Health Effects" In *The Impact of Pesticides,* edited by M. Jokanovic, AcademyPublish.org, UK, 21-38.

[13] Klotz, D. M., Arnold, S. F., McLachlan, J. A., 1997. Inhibition of 17 beta-estradiol and progesterone activity in human breast and endometrial cancer cells by carbamate insecticides, *Life Sci.* 60: 1467-1475.

[14] Dhouib, I. B., Annabi, A., Jallouli, M., Marzouki, S., Gharbi, N., Elfazaa, S., Lasram, M. M., 2016. Carbamates pesticides induced immunotoxicity and carcinogenicity in human: a review, *J. Appl. Biomed.* 14: 85-90.

[15] Lin, C. M., Wei, L. Y., and Wang, T. C., 2007. The delayed genotoxic effect of N-nitroso N-propoxur insecticide in mammalian cells, *Food Chem. Toxicol.* 45: 928-934.

[16] Lemley, A. T., Zhong, W. Z., Janauer, J. E., Rossi, R., 1984. "Investigation of degradation rates of carbamate pesticides: exploring a new detoxification method" In *Treatment and Disposal of Pesticide Wastes*, edited by R. F. Krueger and J. N. Seiber, ACS publications, Washington DC, US, 245-259.

[17] Union Carbide, *Temik-Aldicarb pesticide removal of residues from water. Union Carbide Agricultural Company (internal publication),* Texas, US, 1979.

[18] Yang, G. P., Zhao, Y. H., Lu, X.L., Gao, X. C., 2005. Adsorption of methomyl on marine sediments, *Colloids Surf. A: Physicochem. Eng. Asp.* 264: 179-186.

[19] El-Geundi, M. S., Nassar, M. M, Farrag, T. E., Ahmed, M. H., 2013. Methomyl adsorption onto Cotton Stalks Activated Carbon (CSAC): equilibrium and process design, *Procedia Environ. Sci.* 17: 630-639.

[20] Ahmad, K. S., 2020. Fungicidal Methyl-2-Benzimidazole carbamate adsorption in soil and remediation via Prunus Dulcis derived activated carbon, *Rev. Int. Contam. Ambient* 36: 429-442.

[21] Beesley, L., Moreno-Jimenez, E., Gomez-Eyles, J. L., Harris, E., Robinson, B., Sizmur, T., 2011. A review of biochars' potential role in the remediation, revegetation and restoration of contaminated soils, *Environ. Pollut.* 159: 3269-3282.

[22] Yu, X. Y., Ying, G. G., Kookana, R. S., 2009. Reduced plant uptake of pesticides with biochar additions to soil, *Chemosphere* 76: 665-671.

[23] Zhang, P., Sun, H., Yu, L., Sun, T., 2013. Adsorption and catalytic hydrolysis of carbaryl and atrazine on pig manure-derived biochars: Impact of structural properties of biochars, *J. Hazard. Mater.* 244-245: 217-224.

[24] Chen, J., Hu, Z., Ji, R., 2012. Removal of carbofuran from aqueous solution by orange peel, *Desalination Water Treat.* 49: 106-114.

[25] Chattoraj, S., Mondal, N. K., Das, B., Roy, P., Sadhukhan, B., 2013. Biosorption of carbaryl from aqueous solution onto Pistia stratiotes biomass, *Appl. Water Sci.* 4: 79-88.

[26] Boudesocque, S., Guillon, E., Aplincourt, M., Martel, F., Noael, S., 2008. Use of a low cost biosorbent to remove pesticides from waste water, *J. Environ. Qual.* 37: 631-638.

[27] Zimmer, T. R., Silva J. M., Rocha D. H. de A., Teles H. L., Barbosa D. S., Freitas F. F., Seolatto A. A., 2020. Removal of the pesticide methomyl contained in simulated effluent from equipment washing by adsorption in residual orange bagasse, *Res. Soc. Dev.* 9: e3569118528.

[28] Guiza, S., 2017. Biosorption of heavy metal from aqueous solution using cellulosic waste orange peel, *Ecolog. Eng.* 99: 134-140.

[29] Gisi, S. D., Lofrano, G., Grassi, M., Notarnicola, M., 2016. Characteris-tics and adsorption capacities of low-cost sorbents for wastewater treatment: A review, *Sustain. Mater. Technol.* 9: 10-40.

[30] Latrous El Atrache, L., Hachani, M., Kefi, B. B., 2015. Carbon nanotubes as solid-phase extraction sorbents for the extraction of carbamate

insecticides from environmental waters, *Int. J. Environ. Sci. Technol.* 13: 201-208.

[31] Derbalaha, A., El-Safty, S. A., Shenashena, M. A., Abdel Ghanya, N. A., 2015. Mesocage collector cavities as nanopockets for remediation and real assessment of carbamate pesticides in aquatic water, *Nano-Struct. Nano-Objects* 3: 17-27.

[32] Mason, Y. Z., Choshen, E., Rav-Acha, C., 1990. Carbamate insecticides: Removal from water by chlorination and ozonation, *Water Res.* 24: 11-21.

[33] UNFAO (Food and Agriculture Organization of the United Nations), WHO (World Health Organization), Methiocarb. *Joint Meeting of the FAO Panel of Experts on Pesticide Residues in Food and the Environment and the WHO Core Assessment Group,* Rome, 1999, 132: 531-601.

[34] Tian, F., Qiang, Z., Liu, W., Ling, W., 2013. Methiocarb degradation by free chlorine in water treatment: Kinetics and pathways, *Chem. Eng. J.* 232: 10-16.

[35] Qiang, Z., Tian, F., Liu, W., Liu, C., 2014. Degradation of methiocarb by monochloramine in water treatment: Kinetics and pathways, *Water Res.* 50: 237-244.

[36] Greyshock, A. E., Vikesland, P. J., 2006. Triclosan Reactivity in Chloraminated Waters, *Environ. Sci. Technol.* 40: 2615-2622.

[37] Cruz-Alcalde, A., Sans, C., Esplugas S., 2017. Exploring ozonation as treatment alternative for methiocarb and formed transformation products abatement, *Chemosphere* 186: 725-732.

[38] Li, M. K., Wang, C., Yau, M. L., Bolton, J. R., Qiang, Z. M., 2017. Sulfamethazine degradation in water by the VUV/UV process: Kinetics, mechanism and antibacterial activity determination based on a mini-fluidic VUV/UV photoreaction system, *Water Res.* 108: 348-355.

[39] Yang, L., Li, M., Li, W., Bolton, J. R., Qiang, Z., 2018. A Green Method to Determine VUV (185 nm) Fluence Rate Based on Hydrogen Peroxide Production in Aqueous Solution, *Photochem. Photobiol.* 94: 821-824.

[40] Fang, J., Fu, Y., Shang, C., 2014. The Roles of Reactive Species in Micropollutant Degradation in the UV/Free Chlorine System, *Environ. Sci. Technol.* 48: 1859-1868.

[41] Dong, H., Qiang, Z., Hu, J., Qu, J., 2017. Degradation of chloramphenicol by UV/chlorine treatment: Kinetics, mechanism and enhanced formation of halonitromethanes, *Water Res.* 121: 178-185.

[42] Sweeny, K. H., 1981. "The reductive treatment of industrial wastewater: II Process applications" In *American Institute of Chemical Engineers Symposium Series*, edited by G. F. Bennett, American Institute of Chemical Engineers, New York, Vol. 77, 209: 72-78.

[43] Matheson, L. J., Tratnyek, P. G., 1994. Reductive dehalogenation of chlorinated methanes by iron metal, *Environ. Sci. Technol.* 28: 2045-2053.

[44] Ghauch, A., Gallet, C., Charef, A., Rima, J., Martin-Bouyer, M., 2001. Reductive degradation of carbaryl in water by Zero-valent iron, *Chemosphere* 42: 419-424.

[45] Fenton, H. J. H., 1894. Oxidation of tartaric acid in the presence of iron, *J. Chem. Soc.* 65: 899-910.

[46] Haber, F., and Weiss, J. J., 1934. The catalytic decomposition of H_2O_2 by iron salts, *Proc. R. Soc. Lond. A* 147: 332-351.

[47] Zhang, M., Dong, H., Zhao, L., Wang, D., Meng, D., 2019. A review on Fenton process for organic wastewater treatment based on optimization perspective, *Sci. Total Environ.* 670: 110-121.

[48] Watts, R. J., Bottenberg, B. C., Jensen, M. D., Hess, T. H., Teel, A. L., 1999. Mechanism of the Enhanced Treatment of Chloroaliphatic Compounds by Fenton-Like Reactions, *Environ. Sci. Technol.* 33: 3432-3437.

[49] Babuponnusami, A., Muthukumar, K., 2014. A review on Fenton and improvements to the Fenton process for wastewater treatment, *J. Environ. Chem. Eng.* 2: 557-572.

[50] Neyens, E., and Baeyens, J., 2003. A review of classic Fenton's peroxidation as an advanced oxidation technique, *J. Hazard. Mater.* 98: 33-50.

[51] Poulopoulos, S. G., Nikolaki, M., Karampetsos, D., Philippopoulos, C. J., 2008. Photochemical treatment of 2-chlorophenol aqueous solutions using ultraviolet radiation, hydrogen peroxide and photo-Fenton reaction, *J. Hazard. Mater.* 153: 582-587.

[52] Kuo, W. S., Wu, L. N., 2010. Fenton degradation of 4-chlorophenol contaminated water promoted by solar irradiation, *Sol. Energy* 84: 59-65.

[53] Liu, J., Wu, J., Kang, C., Peng, F., Liu, H., Yang, T., Shi, L., Wang, H., 2013. Photo-Fenton effect of 4-chlorophenol in ice, *J. Hazard. Mater.* 261: 500-511.

[54] Pliego, G., Xekoukoulotakis, N., Venieri, D., Zazo, J. A., Casas, J. A., Rodriguez, J. J., Mantzavinos, D., 2014. Complete degradation of the persistent antidepressant sertraline in aqueous solution by solar photo-Fenton oxidation, *J. Chem. Technol. Biotechnol.* 89: 814-818.

[55] Jo, W., Tayade, R. J., 2014. New generation energy-efficient light source for photocatalysis: LEDs for environmental applications, *Ind. Eng. Chem. Res.* 53: 2073-2084.

[56] Huston, P. L., Pignatello, J. J., 1999. Degradation of selected pesticide active ingredients and commercial formulations in water by the photo-assisted Fenton reaction, *Water Res.* 33: 1238-1246.

[57] Oturan, N., Oturan, M. A., 2018. "Electro-Fenton Process: Background, New Developments, and Applications" In *Electrochemical Water and Wastewater Treatment*, edited by C. A. M. Huitle, M. A. Rodrigo and O. Scialdone, Elsevier, Amsterdam, 193-221.

[58] Sires, I., Brillas, E., Oturan, M. A., Rodrigo, M. A., Panizza, M., 2014. Electrochemical advanced oxidation processes: today and tomorrow. A review, *Environ. Sci. Pollut. Res.* 21: 8336-8367.

[59] Celebi, M. S., Oturan, N., Zazou, H., Hamdani, M., Oturan, M. A., 2015. Electrochemical oxidation of carbaryl on platinum and boron-doped diamond anodes using electro-Fenton technology, *Sep. Purif. Technol.* 156: 996-1002.

[60] Selva, T. M. G., Paixao, T. R. L. C., 2016. Electrochemical quantification of propoxur using a boron-doped diamond electrode, *Diam. Relat. Mater.* 66: 113-118.

[61] Saltmiras, D. A., Lemley, A. T., 2001. Anodic Fenton Treatment of Treflan MTF, *J. Environ. Sci. Health A Tox. Hazard. Subst. Environ. Eng.* 36: 261-274.

[62] Wang, Q. Q., Lemley, A. T., 2003. Competitive Degradation and Detoxification of Carbamate Insecticides by Membrane Anodic Fenton Treatment, *J. Agric. Food Chem.* 51: 5382-5390.

[63] Wang, Q. Q., Lemley, A. T., 2003. Oxidative Degradation and Detoxification of Aqueous Carbofuran by Membrane Anodic Fenton Treatment, *J. Hazard. Mater.* 98: 241-255.

[64] Knepper, T. P., Petrovic, M., de Voogt, P., 2003. "Occurrence of surfactants in surface waters and freshwater sediments: Alkylphenol ethoxylates and their degradation products" In *Analysis and fate of surfactants in aquatic environment*, edited by T. P. Knepper, D. Barcelo and W.P. de Voogt, Elsevier, Amsterdam, *Comprehensive Anal. Chem.* Vol. 40: 675-694.

[65] Cho, Y., Park, H., Choi, W., 2004. Novel complexation between ferric ions and nonionic surfactants (Brij) and its visible light activity for CCl_4 degradation in aqueous micellar solutions, *J. Photochem. Photobiol. A* 165: 43-50.

[66] Prevot, A. B., Pramauro, E., de la Guardian, M., 1999. Photocatalytic degradation of carbaryl in aqueous TiO_2 suspensions containing surfactants, *Chemosphere* 39: 493-502.

[67] Barrios, N., Sivov, P., D'andrea, D., Nunez, O., 2005. Conditions for selective photocatalytic degradation of naphthalene in Triton X-100 water solutions, *Int. J. Chem. Kinet.* 37: 414-419.

[68] Johnson, S., Dureja, P., 2002. Effect of surfactants on persistence of AZADIRACHTIN-A (Neem based Pesticide), *J. Environ. Sci. Health B* 37: 75-80.

[69] Kong, L., Lemley, A. T., 2007. Effect of nonionic surfactants on the oxidation of carbaryl by anodic Fenton treatment, *Water Res.* 41: 2794-2802.

[70] Ye, P., Lemley, A. T., 2008. Adsorption effect on the degradation of Carbaryl, Mecoprop, and Paraquat by anodic Fenton treatment in an SWy-2 Montmorillonite clay slurry, *J. Agric. Food Chem.* 56: 10200-10207.

[71] Romero, A., Santos, A., Cordero, T., Rodriguez-Mirasol, J., Rosas, J. M., Vicente, F., 2011. Soil remediation by Fenton-like process: Phenol removal and soil organic matter modification, *Chem. Eng. J.* 170: 36-43.

[72] Ye, P., Kong, L., Lemley, A. T., 2009. Kinetics of Carbaryl degradation by anodic Fenton treatment in a humic-acid-amended artificial soil slurry, *Water Environ. Res.* 81: 29-39.

[73] Wang, Q. Q., Lemley, A. T., 2004. Kinetic effect of humic acid on Alachlor degradation by anodic Fenton treatment, *J. Environ. Qual.* 33: 2343-2352.

[74] Fan, C., Tsui, L., Liao, M. C., 2011. Parathion degradation and its intermediate formation by Fenton process in neutral environment, *Chemosphere* 82: 229-236.

[75] Agustina, T. E., Ang, H. M., Vareek, V. K., 2005. A review of synergistic effect of photocatalysis and ozonation on wastewater treatment, *J. Photochem. Photobiol. C: Photochem. Rev.* 6: 264-273.

[76] Mehrjouei, M., Muller, S., Moller, D., 2015. A review on photocatalytic ozonation used for the treatment of water and wastewater, *Chem. Eng. J.* 263: 209-219.

[77] Farre, M. J., Franch, M. I., Ayllon, J. A., Peral, J., Domenech, X., 2007. Biodegradability of treated aqueous solutions of biorecalcitrant pesticides by means of photocatalytic ozonation, *Desalination* 211: 22-33.

[78] Rajeswari, R., Kanmani, S., 2009. A study on synergistic effect of photocatalytic ozonation for carbaryl degradation, *Desalination* 242: 277-285.

[79] Farre, M. J., Franch, M. I., Malato, S., Ayllon, J. A., Peral, J., Domenech, X., 2005. Degradation of some biorecalcitrant pesticides by homogeneous and heterogeneous photocatalytic ozonation, *Chemosphere* 58: 1127-1133.

[80] Tamimi, M., Qourzal, S., Assabbane, A., Chovelon, J. M., Ferronato, C., Ait-Ichou, Y., 2006. Photocatalytic degradation of pesticide methomyl: determination of the reaction pathway and identification of intermediate products, *Photochem. Photobiol. Sci.* 5: 477-482.

[81] Tennakone, K., Tilakaratne, C. T. K., Kottegoda, I. R. M., 1997. Photomineralization of carbofuran by TiO_2-supported catalyst, *Water Res.* 31: 1909-1912.

[82] Kanan, S., Moyet, M. A., Arthur, R. B., Patterson, H. H., 2019. Recent advances on TiO_2-based photocatalysts toward the degradation of pesticides and major organic pollutants from water bodies, *Catal. Rev.* 62: 1-65.

[83] Ong, C. B., Ng, L. Y., Mohammad, A. W., 2018. A review of ZnO nanoparticles as solar photocatalysts: Synthesis, mechanisms and applications, *Renew. Sust. Energy Rev.* 81: 536-551.

[84] Daneshvar, N., Aber, S., Seyeddorraji, M., Khataee, A., Rasoulifard, M., 2007. Photocatalytic degradation of the insecticide diazinon in the presence of prepared nanocrystalline ZnO powders under irradiation of UV-C light, *Sep. Purif. Technol.* 58: 91-98.

[85] Fenoll, J., Hellin, P., Flores, P., Martínez, C. M., Navarro, S., 2013. Degradation intermediates and reaction pathway of carbofuran in leaching water using TiO_2 and ZnO as photocatalyst under natural sunlight, *J. Photochem. Photobiol. A* 251: 33-40.

[86] Lhomme, L., Brosillon, S., Wolbert, D., 2008. Photocatalytic degradation of pesticides in pure water and a commercial agricultural solution on TiO_2 coated media, *Chemosphere* 70: 381-386.

[87] Barakat, N. A. M., Nassar, M. M., Farrag, T. E., Mahmoud, M. S., 2013. Effective photodegradation of methomyl pesticide in concentrated solutions by novel enhancement of the photocatalytic activity of TiO_2 using $CdSO_4$ nanoparticles, *Environ. Sci. Pollut. Res.* 21: 1425-1435.

[88] Yang, B., Meng, J., Zhang, Y., Liu, C., Gan, J., Shu, J., 2011. Experimental studies on the heterogeneous reaction of NO_3 radicals with suspended carbaryl particles, *Atmos. Environ.* 45: 2074-2079.

[89] Arias, M., Garcia-Rio, L., Mejuto, J. C., Rodriguez-Dafonte, P., Simal-Gandara, J., 2005. Influence of Micelles on the Basic Degradation of Carbofuran, *J. Agric. Food Chem.* 53: 7172-7178.

[90] Tomasevic, A., Mijin, D., Marinkovic, A., Radisic, M., Prlainovic, N., Durovic-Pejcev, R., Gasic, S., 2017. The photocatalytic degradation of carbofuran and Furadan 35-ST: the influence of inert ingredients, *Environ. Sci. Pollut. Res.* 24: 13808-13822.

[91] Kuo, W. S., Chiang, Y. H., Lai, L. S., 2008. Solar photocatalysis of carbaryl rinsate promoted by dye photosensitization, *Dyes Pigm.* 76: 82-87.

[92] Galadi, A.; Julliard, M., 1996. Photosensitized oxidative degradation of pesticides, *Chemosphere* 33: 1-15.

[93] Bhatt, P., Bhatt, K., Sharma, A., Zhang, W., Mishra, S., and Chen, S., 2021. Biotechnological basis of microbial consortia for the removal of pesticides from the environment, *Crit. Rev. Biotechnol.* 41: 317-338.

[94] Kim, H., Kim, D. U., Lee, H., Yun, J., Ka, J. O., 2017. Syntrophic biodegradation of propoxur by *Pseudaminobacter* sp. SP1a and *Nocardioides* sp. SP1b isolated from agricultural soil, *Int. Biodeterior. Biodegrad.* 118: 1-9.

[95] Mishra, S., Pang, S., Zhang, W., Lin, Z., Bhatt, P., Chen, S., 2021. Insights into the microbial degradation and biochemical mechanisms of carbamates, *Chemosphere* 279: 130500.

[96] El-Ansary, M. S. M., Hamouda, R. A. F., Ahmed-Farid, O. A., 2020. Bioremediation of Oxamyl compounds by Algae: Description and traits of Root-Knot nematode control, *Waste Biomass Valor.* 12: 251-261.

[97] Ufarte, L., Laville, E., Duquesne, S., Morgavi, D., Robe, P., Klopp, C., Potocki-Verronese, G., 2017. Discovery of carbamate degrading enzymes by functional metagenomics, *PloS One* 12: e0189201.

[98] Bhatt, P., Zhou, X., Huang, Y., Zhang, W., Chen, S., 2021. Characterization of the role of esterases in the biodegradation of organophosphate, carbamate, and pyrethroid pesticides. *J. Hazard. Mater.* 411: 125026.

[99] Bhatt, P., Joshi, T., Bhatt, K., Zhang, W., Huang, Y., Chen, S., 2021. Binding interaction of glyphosate with glyphosate oxidoreductase and C-P lyase: Molecular docking and molecular dynamics simulation studies, *J. Hazard. Mater.* 409: 124927.

Chapter 3

Microbial Degradation of Carbamate Pesticides

Elif Aydın*
Disinfection, Sterilization and Antisepsis Program, Kütahya Health Sciences University, Kütahya, Turkey

Introduction

After World War II, the production and use of synthetic pesticide chemicals increased significantly. This increase aroused the curiosity of scientists. Therefore, research began. The early research was concerned with the degradation processes. After the discovery of the ability of microorganisms to degrade xenobiotics, studies were conducted to determine microbial metabolism. As a result of these studies, it was found that microorganisms can degrade any synthetic molecule or chemical which led to a better understanding of the enzymology and biochemistry of pesticide metabolism [1].

As a result of these studies and the finding of their broad biological activity, easy decomposition, and lower persistence, carbamates have been used insecticides, herbicides, fungicides, nematicides, molluscicides, and acaricides in recent agriculture as an alternative to organochlorine pesticides (OC) [1-2].

Carbamates have both advantages and disadvantages. The toxic residues left in the environment by the overuse of these compounds can cause severe toxicity in living organisms. This toxic effect is due to their inhibitory activities against the enzyme acetylcholinesterase. Carbamates are reversible

* Corresponding Author's Email: elif.aydin@ksbu.edu.tr.

In: The Science of Carbamates
Editor: Güllü Kaymak
ISBN: 978-1-68507-708-2

inhibitors of the enzyme acetylcholine (ACh). The enzyme acetylcholinesterase is very important for the necessity of neurotransmission signals in living organisms. It is an essential enzyme involved in the transmission of nerve signals to skeletal muscles and tissues. Inhibition of AChE impairs the function of the neurotransmitter acetylcholine in synapses and neuromuscular junctions in the brain, leading to accumulation of ACh and thus toxic effects.

Carbamates have a cytotoxic effect on individual and animal cells. Laboratory studies show carbamates have a cytotoxic effect on hamster oocytes and inhibit cell viability [3]. It can negatively affect the reproduction, kidneys, liver, nervous system, and immune system of bees, fish, birds, mammals, and marine organisms [4-6]. High intensity induces changes in glucose, proteins, alkaline phosphatases, hemoglobin, serum lipid, malondialdehyde (MDA), and glutathione (GSH) levels in mammalian organisms [7-8].

Carbamates can significantly induce apoptosis and necrosis in human natural killer (NK) cells, and cause apoptosis in human immune cells and lymphocytes [9]. Other effects of carbamates include induction of oxidative stress, inhibition of erythrocyte enzymatic activity, lipid peroxidation, and disruption of cell membrane integrity [10]. Exposure to these compounds for a prolonged period causes behavioral abnormalities and decreased growth rate in bees, birds, and aquatic animals [11].

Carbamate residue is a serious threat to living things due to its toxic effects. Therefore, their elimination is a global priority. However, many conventional methods for the removal of carbamate residues are often not used because they are expensive and cause post-treatment contamination. For these reasons, microbial technology replaces traditional methods for the remediation of carbamate-contaminated areas [12-13].

Microbial Technology in Carbamates

Microorganisms can biodegrade a wide range of environmental contaminants, including carbamate insecticides. Many factors such as microbial degradation, physicochemical mechanisms, photochemical mechanisms, and bioremediation play an important role in the biodegradation of pesticides [14-15].

The long-standing use of pesticides has led many microorganisms to evolve mechanisms to degrade toxic compounds through pesticides, with various mechanisms and enzymatic pathways. Organophosphate-degrading bacteria were originally isolated in a rice field in the Philippines in 1973. Later, many phylogenetic bacteria have been isolated that can metabolize pesticides and use these metabolites as nutrients and carbon sources [16-17].

Microbial technology is an important tool for the removal of carbamate pesticides from contaminated environments by increasing the metabolic activity of microorganisms and their biodegradation processes. Due to this property, it is considered to be environmentally friendly. Microbial degradation is a primary mechanical healing process that effectively breaks down and detoxifies organic contaminants into small, simple molecules. Microbial degradation is a natural, safe, clean, and less destructive method to the environment as it acts as a cleanser. These methods completely reduce pollutants to water, carbon dioxide, or less toxic forms and reduce the need to destroy polluted materials. Moreover, cost and survivability in adverse environmental conditions make them practical and effective [14-15].

The major source of microorganisms capable of degrading carbamates is soil. Other microbial sources include effluents and sediments from the carbamate industry, activated sludge, surface water, and microorganisms from sediments. Many groups of microorganisms have been isolated that differ in their proliferation and degradation capacity of insecticides.

Microorganism species that cause carbamate degradation [18-19]:

- In the group of bacteria, there are Agrobacterium, Athrobacter, Bacillus, Clostridium, Corynebacterium, Flavobacterium, Klebsiella, Pseudomonas, and Xanthomonas
- In the group of fungi, there are Alternaria, Aspergillus, Cladosporium, Fusarium, Glomerella, Mucor, Penicillium, Rhizoctonia, and Trichoderma
- In the group of Actinomycetes, there are Micromonospora, Nocardia, and Streptomyces.

Microbial degradation of carbamate compounds occurs by hydrolysis and oxidation. The enzyme-based degradation method has the potential for rapid action. The few enzymes that hydrolyze carbamate compounds are either esterase or amides. The hydrolysis of the carbamate compound is affected by the biochemical forms of the side chain and the substrate. The parental compound is usually purified by this hydrolytic evolution. The first pathway hydrolysis yields carbon dioxide along with alcohol and methylamine. This pathway is catalyzed by carboxyl ester hydrolases (CBE). These enzymes belong to the group of ester hydrolases and have an important function in carbamate degradation. MCD and CehA are carboxylesterase enzymes most involved in carbamate degradation [20-21].

Microbial carbamate hydrolase genes (cehA, cahA, cfdJ, mcbA, mcd) are important for the development and function of carbamate hydrolase enzymes. However, the precise localization and relevant uses of these enzymes are poorly understood [22].

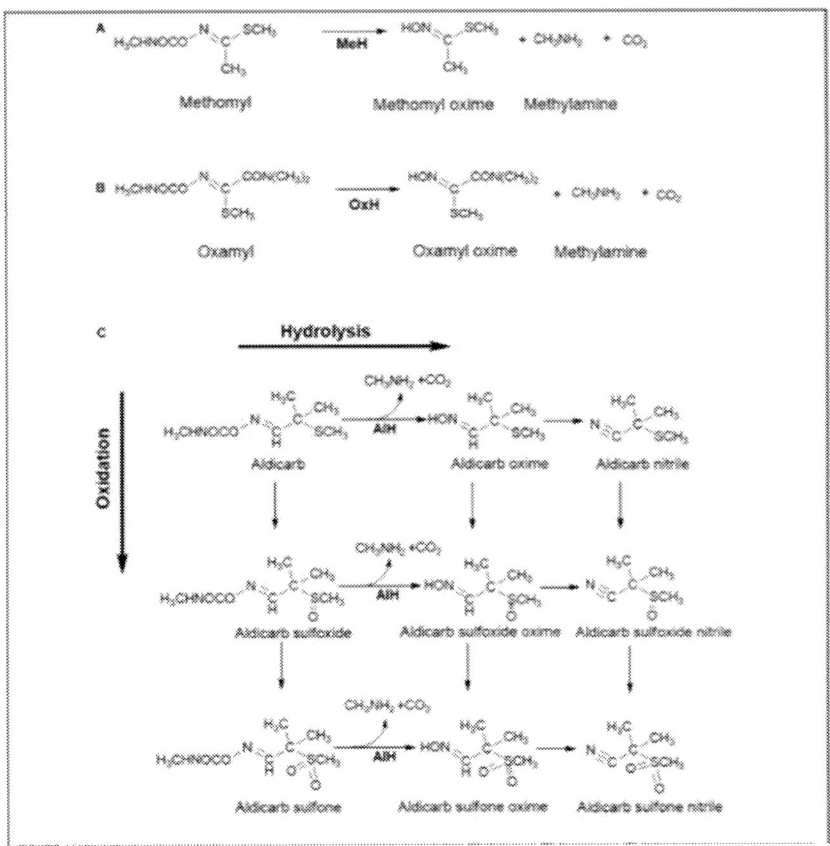

Figure 1. Metabolic steps involved in the oxime-based carbamate pesticide degradation [25]. (A) Methomyl; (B) Oxamyl; and (C) Aldicarb. Various enzymes described for the biological transformation are Methomyl hydrolase, MeH; Oxamyl hydrolase, OxH; Aldicarb hydrolase, AlH. The metabolism of aldicarb involves the hydrolytic pathway leading to the formation of oximes, and after additional dehydration to nitriles, the oxidative pathway leads to the formation of sulfones via sulfoxides. Also, the metabolic pathways for these insecticides are not explained in detail.

Bacterial plasmids play an important role in the evolution and proliferation of microbial degradation genes. These genes have conferred microorganisms with the ability to degrade environmental pollutants. Plasmid-encoded catabolic genes have been detected in only a few species from the genera Achromobacter, Pseudomonas, and Sphingomonas spp. Plasmids CF01, CF02, CF03, CF04 CF05, and CF06 reduce the toxic effect of carbofuran through plasmid-mediated mineralization [23].

In Arthrobacter spp. strain RC100 plays an important role in the complete decomposition of carbaryl compounds. This strain carries three plasmids, two of which are conjugative (pRC1 and pRC2) and one is non-conjugative (pRC300) [24].

Plasmid pAC200 in Rhizobium spp. carries the cehA gene, which encodes carbaryl hydrolase activity. The AC100 plasmid also plays a role in degradation [26].

Plasmid pXWY-1 from *Pseudomonas putida* which carries a gene cluster, and 13 genes in this gene group are directly fully involved in carbaryl degradation [27].

Various microorganisms distinguished and defined for effective degradation of carbamates are listed in Table 1.

Table 1. Microbial degradation of various carbamate pesticides by isolated microorganisms [28]

Pesticides	*Microbes*	*Environmental media*	*Carbamates used as a source of carbon or nitrogen or both*
Carbofuran	*Sphingomonas* sp.	Agriculture soil	C-N-S *
	Novosphingobium sp.	Carbofuran contaminated sludge	C-S **
	Paracoccus sp.	Carbofuran contaminated sludge	C-S **
	Enriched bacterial consortium	Soil	C-S **
	Burkholderia sp.	Rhizospheric soil of rice	C-S **

Table 1. (Continued)

Pesticides	*Microbes*	*Environmental media*	*Carbamates used as a source of carbon or nitrogen or both*
	Novosphingobium sp. KN65.2	Carbofuran exposed-vietanamese soil	C-N-S *
	E. cloacae TA7	Soil	C-N-S *
	Upriavidus sp. ISTL7	Land-fills	C-S **
	Enterobacter sp.	Carbofuran contaminated soil	C-S **
	Sphingbium sp. CFD-1	Activated sludge sample	C-S **
	M. ramannianus	Soil	Not specified
	Trametes versicolor	–	Not specified
Carbaryl	*Arthrobacter* sp. RC 100	Soil	C-S **
	Burkholderia sp. C3	Petroleum contaminate soil	C-S **
	Rhizobium sp. AC100	Carbaryl-contaminated soil	C-S **
	Pseudomonas sp. strains C4, C5, C6	–	C-S **
	Pseudomonas sp. C7	Sediments	C-S **
	Pseudomonas sp. XWY-1	Activated sludge	C-S **
	Micrococcus sp.	Soil	C-S **
	E. cloaceae TA7	Soil	C-N-S *
	Paecilomyces sp.	Soil	C-S **
	Stenotrophomonas sp. YC-1		Not specified
	R. capsulata	Soil	Not specified

Pesticides	*Microbes*	*Environmental media*	*Carbamates used as a source of carbon or nitrogen or both*
	R. sphaeroides	Soil	Not specified
Propoxur	*Arthrobacter* sp.	Soil	C-S **
	Pseudomonas sp.	Soil	C-N-S *
	N. subflava	Soil	C-S **
Methomyl	*E. coli*	Soil	C-S **
	P. aeruginosa	Soil	C-S **
	Stenotrophomonas sp.	Soil	C-S **
	B.cereus *B. safensis*	Crop field	C-S **
	Aminobacter sp. MDW-2 *Afipia* sp. MDW-3	Wastewater sludge	C-S **
	Rhodococcus kroppenstedtii AHJ 3	Rhizospheric soil	Not specified
Oxamyl	*Pseudomonas* sp. OXA20	Soil	C-N-S *
	M. luteus	Egyptian soil	C-S **
	A. aminovorans, *C.heintzii,* *A.nigataensis* *Mesorhizobium* sp	Soil	C-S **
	Trichoderma sp.	Soil	C-N-S *
Aldicarb	*E. cloaceae* TA7	Soil	C-N-S *
	Trametes versicolor	–	Not specified
	A. paraoxydans *P.putida*	Cotton field soil	C-S **
	C.elegans, *P.multicolor* *T. harzianum*	–	C-S **
	S. maltophilia	Soil	C-S **

*C-N-S:Carbon, Nitrogen, Source ** C-S: Carbon, Source.

Carbofuran (2,3-Dihydro-2,2-Dimethyl-7-Benzofuranyl N-Methyl Carbamate)

Carbofuran is a toxic, nematicidal, insecticidal, and acaricidal carbamate with a broad spectrum of activity. This compound is widely used in agriculture to control soil-dwelling and leaf-feeding insects. It is extremely toxic because it is a compound that inhibits acetylcholinesterase and butyrylcholinesterase [29-30].

Carbofuran can cause mutagenic, genotoxic, embryotoxic, reproductive toxicity, cytotoxic, endocrine disruption, and skin problems in humans [31].

Carbaryl (1-Naphthyl Methyl Carbamate)

Carbaryl is a broad-spectrum insecticide with moderate toxicity, that is widely used in agriculture and as an ectoparasite for animals. Carbaryl can seriously damage and disrupt the endocrine system. The toxicity of this compound is due to the ester bond that exists between 1-naphthol and N-methyl carbamic acid [32].

Propoxur (2-Isopropoxyphenyl-N-Methylcarbamate)

Propoxur can be defined as a synthetic and non-systemic carbamate with long-lasting effects on insects. This carbamate is commonly used as a home and garden insecticide. Excessive use of propoxur can affect human health and has a carcinogenic effect. Propoxur can be degraded by microorganisms in soil and water system, and this degradation occurs more rapidly in an alkaline environment. These microorganisms are *N. subflava*, Pseudomonas sp. and Arthrobacter sp. [33-34].

Oxamyl (N,N-Dimethyl-2-Methylcaramolyoxyimino-2-(Methylthio) Acetamide)

Oxamyl is a nematicide and insecticide used to control plant nematodes and potato cyst forming nematodes (Globodera sp.) potato crops. Oxamyl readily leaches into aquatic systems due to its rapid mobility and is poorly and sparingly absorbed by soil particles. The primary degradation mechanism of

Oxamyl is hydrolysis. Carbamoyl and oxime fragments are released. It occurs more rapidly in alkaline soils than in neutral soils [35].

Metomyl (S-Methyl-1-N[Methylcarba-Moyl-Oxy]-Thioacetimidate)

Methomyl is the commonly used carbamate and can be implemented to kill eggs, larvae, and adults of various insect species. Direct or long-term exposure to methylmercury causes dermal toxicity, cytotoxicity, neural dysfunction, hepatotoxicity, developmental defects, and serious toxicological problems [36].

Methomyl is degraded by microorganisms into methomyloxime and methylcarbamic acid, which then decomposes into methylamine, CO_2, and H_2O [37].

Aldicarb (2-Methyl-2(Methylthio) Propinaldehyde 0-(Methylcarbamoyl) Oxime)

Aldicarb is a systemic and comprehensive insecticide belonging to the oxime class. Aldicarb is used to control early-season insects and plant-parasitic nematodes.

Aldicarb is a carbamate that is relatively soluble in water and soil, can be easily absorbed by plants' roots, and migrates along the plant stem. These carbamates can migrate along hydrogeologic flow paths. For this reason, it can contaminate lakes, ponds, rivers, and groundwater resources [38].

Microbial degradation of aldicarb has been studied at contaminated sites but is poorly understood. Bacterial species belonging to *Arthrobacter, Acinetobacter, Enterobacter, Bacillus, Pseudomonas, Methylobacterium,* and *Kocuria* species have been isolated and found to have the ability to utilize aldicarb and its degradation products in the form of nitrogen and carbon [39].

Various techniques are used to detect carbamate degradation pathways in bacteria. These include thin layer chromatography, GC, HPLC, and MS, or a combination include. In addition, the ability to grow, which is the only carbon source on these substances, monitoring of enzyme activity in all cells, the identification of cell-free extract and pathway enzymes are also studied. Whole-cell respiration studies are important for studying oxygen uptake on various substrates and spacers, and for elucidating the factors involved in carbamate degradation. In addition, genomic analyzes identify genes (encoding enzymes) involved in the degradation of carbamates [40].

Fungal Metabolism of Carbamates

Fungal metabolism of carbamates is a developing area, where carbamate pesticides are degraded and/or transformed by various fungal strains belonging to the genera Aspergillus, Mucor, Pichia, Trichoderma, Aschochyta, Trametes, Xylaria, Acremonium, and Gliocladium [41].

In Mucor ramannian, carbofuran, 2-hydroxy-3-(3-methylpropan-2-ol)phenol or 7a(hydroxymethyl) 2,2-dimethylhexahydro-6H-furo[2,3-b] pyran6-one and 3 hydroxycarbofuran-7-phenol is hydrolyzed to carbofuranphenol, which is further metabolized [42]. Pichia anomala degrades the carbofuran, carbaryl and fenobucarb from HQ-C-01. The intermediates in the degradation of carbofuran are benzofuranol, 2-hydroxy-3-(3-methyl propane-2-ol) phenol, and 3-hydroxycarbofuran [43]. Trametes Versicolor, carbofuran, and pyrethroid pesticides degrade imiprothrin and cypermethrin [44]. Aschochyta sp degrades CBS 237.37's mixture of carbofuran, Carbaryl, methiocarb, methomyl, oximil, and propoxitor [45]. Fungal metabolism, like bacterial metabolism, is not fully understood, but the pathways of transformation/mineralization can be studied.

The Genetics and Evolution of Carbamate Decay Processes

The existence and persistence of carbamate pesticides eliminate sensitive microorganisms that they evolve and adapt to the microflora [46]. Genetic variability plays an important role and this process consists of 3 phases:

1. Minor nucleotide changes,
2. Intragenomic mixing and
3. Horizontal gene transfer (HGT)

The third phase allows the organism to obtain unique genetic materials via plasmids, transposons, genomic islands, and integrating conjugative elements [46]. Operons have been identified as genes included in the degradation of some carbamate pesticides. These genes allow for joint cooperation within different genes which complete each other's tasks and are therefore jointly transferred [47].

In addition to HGT, adaptations at the molecular and cellular levels also play an important role in the development and optimization of degradability. At the molecular level, the promiscuity of enzymes is an important factor leading to the evolution of enzymes performing new reactions. Some enzymes can exhibit extensive substrate specificity and act on new compounds to transform them into less toxic intermediates. This non-specific movement reduces the toxicity of the compound, thus increasing the vitality and survival of the organism in a contaminated niche. Because of the advantage they confer to host strains, these enzymes experience trustworthy selection and become effective and specific in their action over a period of time [48]. Among other strategies, compartmentalization of enzymes prevents the interaction of toxic metabolic intermediates with cellular components, thereby increasing degradation efficiency [49].

Conclusion

Carbamates are a serious threat to living things due to their toxic effects. Eliminating these environmental impacts is a global priority. Many of the conventional methods for removing carbamate residues are often too expensive and harmful to be used. For these reasons, microbial technology is replacing traditional methods for remediation of carbamate-contaminated areas.

References

[1] Gupta, R. C. (2006). Classification and uses of organophosphates and carbamates. In *Toxicology of Organophosphate & Carbamate Compounds* (pp. 5-24). Academic Press.

[2] Silberman, J., and Taylor, A. (2020). *Carbamate Toxicity.* Treasure Island, FL: StatPearls Publishing.

[3] Soloneski, S., Kujawski, M., Scuto, A., and Larramendy, M. L. (2015). Carbamates: a study on genotoxic, cytotoxic, and apoptotic effects induced in Chinese hamster ovary (CHO-K1) cells. *Toxicol.* In Vitro 29, 834–844. doi: 10.1016/j.tiv.2015.03. 011.

[4] Dias, E., e Costa, F. G., Morais, S., & de Lourdes Pereira, M. (2015). A review on the assessment of the potential adverse health impacts of carbamate pesticides. *Topics in public health*, 197-212.

[5] Piel, C., Pouchieu, C., Migault, L., Béziat, B., Boulanger, M., Bureau, M., & Baldi, I. (2019). Increased risk of central nervous system tumours with carbamate insecticide use in the prospective cohort AGRICAN. *International journal of epidemiology*, *48*(2), 512-526.

[6] Kaur, R., Mavi, G. K., Raghav, S., & Khan, I. (2019). Pesticides classification and its impact on environment. *Int. J. Curr. Microbiol. Appl. Sci*, *8*(3), 1889-1897.

[7] Morais, S., Dias, E., & Pereira, M. D. L. (2012). Carbamates: human exposure and health effects. *The impact of pesticides*, 21-38.

[8] Sharma, R. K., Jaiswal, S. K., Siddiqi, N. J., & Sharma, B. (2012). Effect of carbofuran on some biochemical indices of human erythrocytes in vitro. *Cellular and Molecular Biology*, *58*(1), 103-109.

[9] Li, Q., Kobayashi, M., and Kawada, T. (2015). Carbamate pesticide-induced apoptosis in human T lymphocytes. *Int. J. Environ. Res. Public Health* 12, 3633–3645. doi: 10.3390/ijerph120403633.

[10] Agarwal S B. A clinical, biochemical, neurobehavioral, and sociopsychological study of 190 patients admitted to hospital as a result of acute organophosphorus poisoning. *Environ. Res.* 1993; 62: 63-70.

[11] Kumar, S., Kaushik, G., Dar, M. A., Nimesh, S., Lopez-Chuken, U. J., & Villarreal-Chiu, J. F. (2018). Microbial degradation of organophosphate pesticides: a review. *Pedosphere*, *28*(2), 190-208.

[12] Feng, Y., Huang, Y., Zhan, H., Bhatt, P., Chen, S., 2020. An overview of strobilurin fungicide degradation: current status and future perspective. *Front. Microbiol.* 11, 389.

[13] Pang, S., Lin, Z., Zhang, W., Mishra, S., Bhatt, P., Chen, S., 2020. Insights into the microbial degradation and biochemical mechanisms of neonicotinoids. *Front. Microbiol.* 11, 868.

[14] Pandey, G., Dorrian, S. J., Russell, R. J., Brearley, C., Kotsonis, S., and Oakeshott, J. G. (2010). Cloning and biochemical characterization of a novel carbendazim (methyl-1H-benzimidazol-2-ylcarbamate)-hydrolyzing esterase from the newly isolated Nocardioides sp. strain SG-4G and its potential for use in enzymatic bioremediation. *Appl. Environ. Microbiol.* 76, 2940–2945. doi: 10.1128/AEM. 02990-09.

[15] Satish, G. P., Ashokrao, D. M., & Arun, S. K. (2017). Microbial degradation of pesticide: A review. *African Journal of Microbiology Research*, 11(24), 992-1012. doi:10.5897/ajmr2016.8402.

[16] Das, A. C., Chakravarty, A., Sen, G., Sukul, P., & Mukherjee, D. (2005). A comparative study on the dissipation and microbial metabolism of

organophosphate and carbamate insecticides in orchaqualf and fluvaquent soils of West Bengal. *Chemosphere,* 58(5), 579-584. doi:10.1016/ j.chemosphere.2004.07.007.

[17] Talwar, M. P., Mulla, S. I., & Ninnekar, H. Z. (2014). Biodegradation of organophosphate pesticide quinalphos by Ochrobactrum sp. strain HZM. *Journal of Applied Microbiology*, 117(5), 1283-1292. doi:10.1111/ jam.12627.

[18] Chaudhry, G. R., & Ali, A. N. (1988). Bacterial metabolism of carbofuran. *Applied and Environmental Microbiology*, 54(6), 1414-1419.

[19] Kim, I. S., Ryu, J. Y., Hur, H. G., Gu, M. B., Kim, S. D., & Shim, J. H. (2004). Sphingomonas sp. strain SB5 degrades carbofuran to a new metabolite by hydrolysis at the furanyl ring. *Journal of Agricultural and Food Chemistry*, 52(8), 2309-2314. doi:10.1021/jf035502l.

[20] Ufarté, L., Laville, E., Duquesne, S., Morgavi, D., Robe, P., Klopp, C., ... & Potocki-Veronese, G. (2017). Discovery of carbamate degrading enzymes by functional metagenomics. *PloS One,* 12(12), 1-21. doi:10.1371/ journal.pone.0189201

[21] Bhatt, P., Zhou, X., Huang, Y., Zhang, W., & Chen, S. (2021). Characterization of the role of esterases in the biodegradation of organophosphate, carbamate, and pyrethroid pesticides. *Journal of Hazardous Materials*, 125026.

[22] Rangasamy, K., Athiappan, M., Devarajan, N., & Parray, J. A. (2017). Emergence of multi drug resistance among soil bacteria exposing to insecticides. *Microbial Pathogenesis*, 105, 153-165. doi:10.1016/j. micpath.2017.02.011

[23] Feng, X., Ou, L. T., & Ogram, A. (1997). Plasmid-mediated mineralization of carbofuran by Sphingomonas sp. strain CF06. *Applied and environmental microbiology*, *63*(4), 1332–1337. https://doi.org/10. 1128/ aem.63.4.1332-1337 .1997.

[24] Hayatsu M, Hirano M, Nagata T. Involvement of two plasmids in the degradation of carbaryl by Arthrobacter sp. strain RC100. *Appl. Environ. Microbiol.* 1999 Mar;65(3):1015-9. doi: 10.1128/AEM.65.3.1015-1019.1999. PMID: 10049857; PMCID: PMC91138.

[25] Malhotra, H., Kaur, S., & Phale, P. S. (2021). Conserved metabolic and evolutionary themes in microbial degradation of carbamate pesticides. *Frontiers in Microbiology*, *12*.

[26] Hashimoto M, Mizutani A, Tago K, Ohnishi-Kameyama M, Shimojo T, Hayatsu M. Cloning and nucleotide sequence of carbaryl hydrolase gene (cahA) from Arthrobacter sp. RC100. *J. Biosci. Bioeng.* 2006 May; 101(5):410-4. doi: 10.1263/jbb.101.410. PMID: 16781470.

[27] Zhu, S., Wang, H., Jiang, W. et al. Genome Analysis of Carbaryl-Degrading Strain Pseudomonas putida XWY-1. *Curr. Microbiol.* 76, 927–929 (2019).

[28] Mishra S, Pang S, Zhang W, Lin Z, Bhatt P, Chen S. Insights into the microbial degradation and biochemical mechanisms of carbamates. Chemosphere. 2021 Sep;279:130500. doi: 10.1016/j.chemosphere. 2021.130500. *Epub* 2021 Apr 7. PMID: 33892453.

[29] Yan, X., Jin, W., Wu, G., Jiang, W., Yang, Z., Ji, J., ... & Hong, Q. (2018). Hydrolase CehA and monooxygenase CfdC are responsible for carbofuran degradation in Sphingomonas sp. strain CDS-1. *Applied and environmental microbiology*, *84*(16), e00805-18.

[30] Li, Z., Wang, X., Ni, Z., Bao, J., Zhang, H., 2020. In-situ remediation of Carbofuran contaminated soil by immobilized white-rot fungi. *Pol. J. Environ. Stud.* 29,1237e1243.

[31] Rocha, O., Gazonato Neto, A. J., Santos Lima, C. J., Freitas, E.C., Miguel, M., de Silva Mansano, A., Moreira, A.R., Daam, M.A., 2018. Sensitivities of three tropical indigenous freshwater invertebrates to single and mixture exposures of diuron and carbofuran and their commercial formulations. *Ecotoxicology* 27, 834e844.

[32] Han, Y., Song, S., Wu, H., Zhang, J., & Ma, E. (2017). Antioxidant enzymes and their role in phoxim and carbaryl stress in Caenorhabditis elegans. *Pesticide biochemistry and physiology*, *138*, 43-50.

[33] Azab, A. E., El-Shafey, A. A., Bayomy, M. F., & Elsayed, A. S. I. (2017). Toxicological Influence of Propoxur on Respiratory Functions of the Blood in Pigeon (Columba livia domestica). *Scj. Acad. J. BioSci.*, *4*, 1093-1100.

[34] Kamini Shetty, D., Trivedi, V. D., Varunjikar, M., Phale, P. S., 2019. Compartmentalization of the carbaryl degradation pathway: molecular characterization of inducible periplasmic carbaryl hydrolase from Pseudomonas spp. *Appl. Environ. Microb*iol. 84, e02115ee02117.

[35] Rousidou, K., Chanika, E., Georgiadou, D., Soueref, E., Katsarou, D., Kolovos, P., Karpouzas, D.G., 2016. Isolation of oxamyl-degrading and identification of cehA as a novel oxamyl hydrolase gene. Front. *Microbiol.* 7, 616.

[36] Saleem, A. A., 2019. Induction of hyper pigmentation and heat shock protein 70 response to the toxicity of methomyl insecticide during the organ development of the Arabian toad, Bufo arabicus (Heyden, 1827). *J. Histotechnol.* 42, 104e115.

[37] Lin, Z., Zhang, W., Pang, S., Huang, Y., Mishra, S., Bhatt, P., Chen, S., 2020. Current approaches to and future perspectives on methomyl degradation in contaminated soil/water environments. *Molecules* 25, 738.

[38] Wang, T., Chamberlain, E., Shi, H., Adams, C. D., Ma, Y., 2011. Comprehensive studies of aldicarb degradation in various oxidantsystems using high performance liquid chromatography coupled with UV detection and quardrupole iron trap mass spectroscopy. *Int. J. Environ. Anal. Chem.* 91, 97e111.

[39] Lawrence, K. S., Feng, Y., Lawrence, G. W., Burmester, C. H., Norwood, S.H., 2005. Accelerated degradation of aldicarb and its metabolites in cotton fields. *J. Nematol.* 37, 190e197.

[40] Mishra, S., Zhang, W., Lin, Z., Pang, S., Huang, Y., Bhatt, P., & Chen, S. (2020). Carbofuran toxicity and its microbial degradation in contaminated environments. *Chemosphere*, *259*, 127419.

[41] Mustapha, M. U., Halimoon, N., Johar, W. L. W., & Abd Shukor, M. Y. (2019). An Overview on Biodegradation of Carbamate Pesticides by Soil Bacteria. *Pertanika Journal of Science & Technology*, *27*(2).

[42] Seo, J., Jeon, J., Kim, S. D., Kang, S., Han, J., Hur, H. G., 2007. Fungal biodegradation of carbofuran and carbofuran phenol by fungus Mucor ramannianus: identification of metabolites. *Water Sci. Technol.* 55, 163e167.

[43] Yang, L., Chen, S., Hu, M., Hao, W., Geng, P., Zhang, Y., 2011. Biodegradation of carbofuran by Pichia anomala strain HQ-C-01 and its application for bioremediation of contaminated soils. *Biol. Fertil. Soils* 47, 917e923.

[44] Mir-Tutusaus, J. A., Masís-Mora, M., Corcellas, C., Eljarrat, E., Barceló, D., Sarrà, M., ... & Rodríguez-Rodríguez, C. E. (2014). Degradation of selected agrochemicals by the white rot fungus *Trametes versicolor*. *Science of the total environment*, 500, 235-242.

[45] Kaur, P., Balomajumdar, C., 2019. Simultaneous biodegradation of mixture of carbamates by newly isolated Ascochyta sp. CBS 237.37. *Ecotoxicol. Environ. Saf.* 167, 590e599.

[46] Phale, P. S., Shah, B. A., Malhotra, H., 2019. Variability in assembly of degradation for operons for naphthalene and its derivative, carbaryl, suggests mobilization through horizontal gene transfer. *Genes* 10, 569.

[47] Rousidou, K. (2020). Isolation of bacteria that degrade carbamate insecticides and characterization of the functional and ecological role of bacterial genes involved in their hydrolysis in soil (Doctoral dissertation, Πανεπιστήμιο Θεσσαλίας. Σχολή Επιστημών Υγείας. Τμήμα Βιοχημείας και Βιοτεχνολογίας. Εργαστήριο Βιοτεχνολογίας Φυτών και Περιβάλλοντος).

[48] Glasner, M. E., Truong, D. P., & Morse, B. C. (2020). How enzyme promiscuity and horizontal gene transfer contribute to metabolic innovation. *The FEBS journal*, 287(7), 1323-1342.

[49] Kamini, Sharma, R., Punekar, N. S., Phale, P. S., 2018. Carbaryl as a carbon and nitrogen source: an inducible methylamine metabolic pathway at the biochemical and molecular level in Pseudomonas sp strain C5pp. *Appl. Environ. Microbiol.* 84, 1866e1918.

Chapter 4

Fungal Biodegradation of Carbamates

Derya Berikten[*]
Training and Research Center, Kütahya Health Sciences University, Kütahya, Turkey

Introduction

Agriculture is a vital part of the world economy, and agriculture's usage of pesticides is commonplace to boost crop yields [1]. Pesticides are also employed in the polymer and pharmaceutical industries [2]. Pesticides have grown pervasive as a result of their widespread use [3] and pose a significant threat to our ecosystems [4]. Carbamates are one of the most commonly used pesticides in agriculture due to their broad range of biological action [5-6]. Carbofuran, carbaryl, aldicarb, methiocarb, and methomyl are generally used carbamates [7-9]. Like organochlorines and organophosphates, carbamates constitute a severe worldwide hazard due to their toxicity and durability; their intermediates and residues frequently build up to dangerous amounts in soil [10]. A significant portion of pesticides used never reach their intended target and frequently caused harm to non-target species. They spread through the soil, water, air, and eventually into humans through food consumption [11]. Several CRBs have already been banned in the United States and the European Union [12] since contaminated soil treatment has become a desirable aim.

Bioremediation is regarded as an ecologically and economically viable method with limitless potential for eliminating toxins (in terms of energy consumption) [13]. The use of bioremediation to remediate pesticide-contaminated soil is advocated as a green and lower-cost alternative to physicochemical treatment [14]. In light of the fact that microbial degradation

[*] Corresponding Author's Email: derya.berikten@ksbu.edu.tr.

In: The Science of Carbamates
Editor: Güllü Kaymak
ISBN: 978-1-68507-708-2

is the most common method for removing pesticides and has huge potential, isolation of pesticide-degrading microorganisms or consortia to boost biodegradation is a pressing necessity. In terms of initial inoculum, the capacity of biological sources to spread is far superior to physical and chemical treatment methods [13]. In soil, microbial degradation is a major pathway for carbamate degradation, although volatilization and photodegradation have negligible impacts [15]. The vast majority of investigations to date have focused only on the isolation of carbamate-degrading bacteria. Unfortunately, there are just a few reports of CRB degradation caused by fungi. Removal of aldicarb, methomyl, and methiocarb by ligninolytic fungus *Trametes versicolor* was investigated by [16] and carbofuran by *Pichia anomala* strain HQ-C-01 [17], *Gliocladium sp.* [18], *Mucor ramannian* [19], a mixture of the carbamates (carbofuran, Carbaryl, methiocarb, methomyl, oxamyl, and propoxur) by *Ascochyta sp.* CBS 237.37 strain [20], carbaryl and carbofuran by *Acremonium sp.* [21], carbaryl, carbofuran, xylylcarb, metolcarb, propoxur, isoprocarb, fenobucarb, and aldicarb by *Aspergillus niger* PY168 strain [22].

Microbial Biodegradation of Pesticides

Chemical treatment, volatilization, and cremation as methods for removing pesticides from soil have been met with public criticism due to issues such as enormous quantities of acids and alkalis created and later disposed of, as well as potentially harmful emissions and increased economic expenditures. The bulk of physical-chemical cleaning solutions are both inefficient and expensive. These clean-up procedures are not suitable for big farms since only tiny soil samples are necessary, and they are carried out in laboratories, requiring a huge amount of resources because polluted soil must be dug at a site and transported to a storage place where it may be stored [23].

The contrast between agricultural goods with high yields or stable production and the damage to the environment is a problem that should be given importance. From one point of view, pesticides with minimal toxicity, high efficacy, and low pesticide remains should be sought for and developed; on the other hand, methods for decomposing pesticide residues should also be given serious consideration. Microbial decomposition of pesticide debris was first researched in the 1940s. As people have become increasingly concerned about the environment, studies on organic contaminants, breakdown processes and mechanisms have become more extensive. Microbial degradation has been increasingly popular in recent years as a result of pesticides being utilized as primary microbial nutrition, eventually decomposing into tiny molecules like CO_2 and H_2O. The process was known as an enzymatic reaction, firstly the compound entering the body of the microorganism, followed by a series of physiological and biochemical reactions involving various enzymes, and finally, the pesticide being completely degraded or broken down into smaller molecular compounds with no or low toxicity [24-25]. Bacteria found in nature might digest pesticide residues at a minimal cost, while also being ecologically beneficial and avoiding secondary contamination. However, the efficiency was modest, and the natural environment was complicated and changing, which might have an impact on the feasibility and efficiency of microbial pesticide breakdown. Detailed research on how organic pesticides are degraded by bacteria has provided more information on this. A variety of microorganisms capable of converting and degrading pesticides have been discovered, among them. Furthermore, the biodegradable ways and processes of pesticides were thoroughly outlined. According to studies, biodegradable pesticides are mostly concentrated in soil microorganisms such as fungus, bacteria, and actinomycetes, with bacteria and fungi playing the most important roles. A more in-depth investigation could be conducted because the bacteria could efficiently produce mutant strains with a range of metabolic capacities to adapt to their environment [26].

Many researchers have recently enriched, isolated, grown, and tested a variety of microorganisms from natural sewage or soil to breakdown pesticides, including bacteria, fungi, actinomycetes, algae, and other microbial strains. Bacteria were isolated from sediments and water samples collected in areas with significant agricultural activity to detect endosulfan degradation [27]. The five bacteria genera klebsiella, acinetobacter, alcaligenes, flavobacterium, and bacillus, were discovered to be capable of degrading endosulfan. Only the seven strains of *Streptomyces alanosinicus, Streptoverticillium album, Nocardia farcinica, Streptomyces atratus, Nocardia vaccini, Nocardia amarae*, and *Micromonospora chalcea* can thrive and breakdown pesticides [28]. Researchers investigated the degradation of atrazine using white-rot fungus, and discovered that the half-life of atrazine had fallen to six days [29]. Furthermore, the potential of the green microalgae *Chlamydomonas mexicana* to digest atrazine was investigated, and discovered that microalgae could efficiently break down atrazine by aggregating it in cells and subsequently decomposing it at a rate of 14–36 percent [30].

Fungal Biodegradation of Pesticides

The handling and disposal of pesticide rinse water created by individual farmers or commercial agrochemical executives have turned into a substantial waste management concern, thanks to more stringent environmental rules governing the clean-up and disposal of large-scale effluent from pesticide industries. Bioremediation systems based on fungi have recently received much interest. Fungi from natural sources can be filtered out as a useful technique for biodegradation of hazardous organic compounds. The fungus can be employed when bacterial species fail to break down chemicals. Fungi, unlike prokaryotes, have exhibited a wide range of metabolic capabilities, resulting in metabolites that are remarkably comparable to those generated by mammalian metabolism. Instead of employing mammalian microsomal fragments or live organisms, this metabolic route indirectly elucidates the metabolic fates of organic molecules occurring in mammalian liver cells. Fungal metabolism also provides a simple approach for producing vast amounts of metabolites [19].

White-rot fungi use their extracellular ligninolytic enzyme system to degrade refractory substances such as xenobiotics and lignin efficiently. When compared to bacteria, the white-rot fungus has a unique advantage. Lignin peroxidase (LiP), Mn-dependent peroxidase (MnP), and laccase make up the ligninolytic enzyme system. The enzyme's H_2O_2 activates the extracellular degradation pathway, producing a variety of free radical chain events that result in no particular oxidative destruction of the substrate. White-rot fungi do not need to be preconditioned to specific contaminants, which means they can break down low-concentration chemicals in the environment to undetectable levels. Furthermore, they create OH-free radicals to alter the pH of the environment, which inhibits microbial invasion [31]. *Phanerochaete* and other similar fungi attack wood using a potent extracellular enzyme called peroxidase, which works on a wide spectrum of chemical compounds. Pesticides, polychlorinated biphenyl, and polycyclic aromatic hydrocarbons are among the substances destroyed by the well-known fungus *Phanerochaete chrysosporium*. It has been claimed that it can break down isoproturon herbicides by solid-state fermentation, which is well recognized for xenobiotic metabolism [32].

Endosulfan degrading aerobic fungi are helpful in the clean-up of organochlorine pesticide-contaminated soil. Bioconversion of endosulfan-affected water bodies and soil may be easily done by *Aspergillus niger*. The endosulfan degrading ability of the fungus *A.niger* was tested. It was discovered that technical grade endosulfan at 400 mg/ml was tolerated by the culture, with the endosulfan completely disappearing after 12 days [33], as designated by the evolution of CO_2. Thin layer chromatography (TLC) analysis demonstrated the development of numerous endosulfan metabolic intermediates, including endosulfan diol, endosulfan sulfate, and an unidentified metabolite. In a separate investigation, ıt was discovered that the white-rot fungus *Trametes hirsuta* successfully degraded endosulfan and endosulfan sulfate. Multiple routes were discovered for the fungus to break down endosulfan and endosulfan sulfate [34]. It has been observed that strains such as *Mortierella* sp. strain W8 and Cm1-45 degraded endosulfan by 50 to 70 percent in 28 days at 25°C via diol production of endosulfan first and then endosulfan lactone, resulting in increased agricultural soil fertility [35].

Under aerobic circumstances, a fungus strain *Fusarium verticillioides* isolated from *Agave tequilana* leaves by enrichment procedures may utilize lindane as a carbon and energy source. The increased degradation efficiency is attained in nitrogen and phosphorus limiting environments. The effectiveness of the biodegradation process was increased by increasing the concentration of lindane and yeast extract and ambient conditions [36-37]. Methomyl and diazinon insecticides are degraded by rot fungus isolated from contaminated soil. The optimal temperature for maximum efficiency is 28°C [38] and the use of mixed fungal strains has the potential to degrade combined pesticides (DDT and chlorpyrifos). The great efficiency of degradation was seen at low concentrations of combined insecticides. The degradation efficiency of DDT and chlorpyrifos was reported to be 26.94 and 24.94 percent, respectively [39]. The phytopathogenic fungus develops quickly on organophosphate herbicides and destroys them quickly as well [40].

Trichoderma viride and *T. harzianum* have great capabilities for degrading pesticides like pirimicarb. The degradation capacity rises when activated charcoal is added [41]. The strain *Sphingomonas yanoikuyaecan* breakdown carbamate and pyrethrin (OPs) with great efficiency under hard circumstances using an enrichment culture technique and was evaluated using gas chromatography [42]. A salt-resistant actinomycete degrades Carbofuran [43]. *S. alanosinicus* is capable of degrading carbofuran by up to 95 percent. It is effective in saline soils because it uses carbofuran as the only source of carbon [44].

The fungus requires chlorpyrifos as its primary source of energy and carbon, causing it to degrade quickly. Basidiomycetes, another fungus, destroy chlorpyrifos extremely well. In liquid batch cultures, fungal strains such as *Fusarium oxysporum, Lentinula edodes, Penicillium brevicompactum,* and

Lecanicillium saksenae have shown great promise in biodegrading pesticides such as difenoconazole, terbuthylazine, and pendimethalin and are being studied as active microorganisms for pesticide degradation [45]. Pesticides may be degraded by more than 30 microorganisms, with the Gliocladium genus having the most activity for selective carbofuran breakdown [46]. Apart from that, different fungi such as *Trametes* sp. and *Polyporus* sp. were shown to be capable of degrading a wide range of substances, including pesticides. The capacity of *A. fumigates, A. sydowii, A. terreus, A. flavus, Fusarium oxysporum*, and *Penicillium chrysogenum* to degrade pesticides has also been observed [47].

Carbamate Degradation by Fungi

Carbamates were first utilized as pesticides in the early 1950s and are still widely used in pest management today because of their efficacy and the broad range of biological action (insecticides, fungicides, and herbicides). Carbamates are used to manage insects and nematodes in soils because of their strong polarity and solubility in water and their thermal instability and severe acute toxicity. Carbamate insecticides are carbamate esters and organic chemicals generated from carbamic acid chemically. Carbamates are AChE inhibitors that are to blame for the bulk of poisonings in rural areas. Several mechanisms in living creatures turn carbamates into diverse compounds, including hydrolysis, bioremediation, oxidation, photolysis, bio-decomposition, and metabolic reactions [48]. Carbamate breakdown by fungi has been documented. A new hydrolase from *Aspergillus niger* capable of hydrolyzing multiple N-methylcarbamate pesticides is of particular interest. CehA has been purified and characterized from *Aspergillus niger* PY168, which can cause hydrolysate carbaryl, carbofuran, xylylcarb, metolcarb, propoxur, isoprocarb, fenobucarb, and aldicarb [22]. Carbofuran, carbaryl, methiocarb, methomyl, oxamyl, and propoxur have all been reported to be degraded by *Aschochyta* sp. CBS 237.37. Carbofuran phenol and 3-hydroxycarbofuran for carbofuran, methomyl oxime for methomyl, and 1-naphthol for carbaryl were discovered as intermediates in the wasted media studies [20].

White rot fungi are a phylogenetic group of fungi capable of degrading lignin. *Phanerochaete, Trametes, Bjerkandera*, and *Pleurotus* are the four primary genera of white-rot fungus showing bioremediation potential. On the other hand, these fungi cannot utilize lignin as an energy source and must instead rely on cellulose or other carbon sources. In addition, fungal growth in a branching, filamentous phase provides for more efficient colonization and exploration of polluted soil. However, the lignin degradation system of enzymes is the predominant biodegradation method used by this group of fungi. Extracellular lignin modifying enzymes (LMEs) have low substrate specificity, allowing them to mineralize a wide spectrum of very resistant organopollutants with lignin-like structures [23]. Carbofuran and the pyrethroid insecticides imiprothrin and cypermethrin have been observed to be degraded by *Trametes versicolor*. The only intermediate discovered in carbofuran breakdown by wasted media analysis was 3-hydroxycarbofuran. The non-specific monooxygenase cytochrome-P450 has been found to have a role in carbofuran breakdown in this strain [49]. Aldicarb, methomyl, methiocarb, and carbofuran have all been found to be transformed by another strain of *T. versicolor*. The action of cytochrome-P450 monooxygenase changed carbofuran into 3-hydroxycarbofuran and 3-ketocarbofuran, but no intermediates were found for methomyl, aldicarb, or methiocarb, however, laccase activity was suggested to be involved [16].

A soil microbe capable of digesting carbamates such as chlorpropham (using soil enrichment techniques) was isolated. They hypothesized that hydrolysis was a primary soil degradation process. The soil microorganisms had varying levels of substrate specificity, but they were all capable of dehalogenating and decomposing a wide range of pesticides (synthesis of 3-chloroaniline and subsequent free chloride ion liberation) [50-51]. Carbofuran, carbaryl, and fenobucarb have all been demonstrated to be degraded by *Pichia anomala* HQ-C-01. Benzofuranol, 2-hydroxy-3-(3-methyl propan-2-ol) phenol, and 3-hydroxycarbofuran were found as carbofuran breakdown intermediates [17].

Mucor ramannianus, a common soil fungus, was used to research the bio-decomposition of carbofuran and carbofuran phenol to understand better the fungus's role in carbofuran and carbofuran phenol degradation. *M. ramannian* hydrolysed carbofuran to carbofuran phenol, which was then metabolized to 2-hydroxy-3-(3-methyl propan-2-ol) phenol or 7a-(hydroxymethyl)-2,2-dimethylhexahydro-6H-furo[2,3-b] pyran-6-one and 3 hydroxycarbofuran-7-phenol [19].

Acremonium and *Xylaria* sp. have different pathways for Carbaryl degradation from the bacteria. Carbaryl esterase, laccase, and cytochrome-P450 are three enzymes involved in the breakdown of carbaryl from *Acremonium* and *Xylaria* sp. Carbaryl and carbofuran have been demonstrated to be degraded by *Acremonium* sp. and used as a carbon and energy source.

Carbofuran-7-phenol and 3-(2-hydroxy-2-methyl propyl) benzene-1,2-diol were formed after the carbofuran was hydrolyzed. Carbaryl decomposition is mediated by 1-naphthol and benzoic acid [21]. *Xylaria* sp. BNL1 has also been shown to be capable of degrading Carbaryl, which was hydrolyzed to 1-naphthol and then degraded to 1,4-naphthoquinone and benzoic acid [52].

Conclusion

Carbamate pesticides are widely employed in agriculture, food, and public health as insecticides, nematicides, acaricides, herbicides, and fungicides. The widespread usage of carbamate insecticides has resulted in their dissemination throughout the environment. This has put pressure on the microbiota to find ways to reduce their toxicity through detoxification, transformation, or mineralisation. Several bacteria and fungi have been found to break down carbamates and use them as a single carbon and nitrogen source. The majority of the existing research on carbamate degradation focuses on their use as a carbon source, with little extensive evaluations on their use as a nitrogen source. The majority of publications on carbamate breakdown focus on the pesticide's bacterial metabolism, whereas the fungal metabolism of carbamates is little studied. Because fungal strains have benefits such as releasing a broad spectrum of extracellular enzymes, tolerance to greater pollutant concentrations, and increased bioavailability, it is important to understand how they work, and more research is needed to do this.

References

[1] Álvarez-Martín, A., Hilton, S. L., Bending, G. D., Rodríguez-Cruza M. S., Sánchez-Martína, M. J. (2016). Changes in activity and structure of the soil microbial community after application of azoxystrobin or pirimicarb and an organic amendment to an agricultural soil. *Appl. Soil Ecol.* 106, 47–57.

[2] Ćwielag-Piasecka, I., Witwicki, M., Jerzykiewicz, M., and Jezierska, J. (2017). Can carbamates undergo radical oxidation in the soil environment? A case study on carbaryl and carbofuran. *Environ. Sci. Technol.* 51 (24), 14124–14134.

[3] Alvarez, A., Saez, J. M., Davila Costa, J. S., Colin, V. L., Fuentes, M. S., Cuozzo, S. A., Benimeli, C. S., Polti, M. A., & Amoroso, M. J. (2017).

Actinobacteria: Current research and perspectives for bioremediation of pesticides and heavy metals. *Chemosphere*, 166, 41–62.

[4] Lizano-Fallas, V., Masís-Mora, M., Espinoza-Villalobos, D., Lizano-Brenes, M., Rodríguez-Rodríguez, C. E. (2017). Removal of pesticides and ecotoxicological changes during the simultaneous treatment of triazines and chlorpyrifos in biomixtures. *Chemosphere*, 182, 106–113.

[5] Trivedi, V. D., Jangir, P., Sharma, R., Phale, P. S. (2016). Insights into functional and evolutionary analysis of carbaryl metabolic pathway from Pseudomonas sp. strain C5pp. *Sci. Rep.* 6, 1–13.

[6] Yang, C., Xu, X., Liu, Y. Jiang, H., Wu, Y., Xu, P., Liu, R. (2017). Simultaneous hydrolysis of carbaryl and chlorpyrifos by Stenotrophomonas sp. strain YC-1 with surface-displayed carbaryl hydrolase. *Sci. Rep.* 7 (1), 1–8.

[7] Lewis, K. A., Tzilivakis, J., Warner, D. J., Green, A. (2016). An international database for pesticide risk assessments and management. *Hum. Ecol. Risk Assess*. 22 (4), 1050–1064.

[8] Kim, I., Kim, D. U., Kim, N. H., Ka, J. O. (2014). Isolation and characterisation of fenobucarb-degrading bacteria from rice paddy soils. *Biodegradation*, 25 (3), 383–394.

[9] Onunga, D. O., Kowino, I. O., Ngigi, A. N., Osogo, A., Orata, F., Getenga, Z. M., Were, H. (2015). Biodegradation of carbofuran in soils within Nzoia River Basin, Kenya. *J. Environ. Sci. Health- Part B Pestic., Food Contam., Agric. Wastes,* 50 (6), 387–397.

[10] Ccanccapa-Cartagena, A., Masiá, A., Picó, Y. (2017). Simultaneous determination of pyrethroids and pyrethrins by dispersive liquid-liquid microextraction and liquid chromatography triple quadrupole mass spectrometry in environmental samples. *Anal. Bioanal. Chem.* 409 (20), 4787–4799.

[11] Gong, T., Xu, X., Dang, Y., Kong, A., Wu, Y., Liang, P., Wang, S., Yu, H., Xu, P., Yang, C. (2018). An engineered Pseudomonas putida can simultaneously degrade organophosphates, pyrethroids and carbamates. *Sci. Total Environ.* 628–629, 1258–1265.

[12] Chin-Pampillo, J. S., Carazo-Rojas, E., Pérez-Rojas, G., Castro-Gutiérrez, V., Rodríguez-Rodríguez, C. E. (2014). Accelerated biodegradation of selected nematicides in tropical crop soils from Costa Rica. *Environ. Sci. Pollut. Res.* 22 (2), 1240–1249.

[13] Ferreira, L., Rosales, E., Sanromán, M. A., Pazos, M. M. (2015). Scale-up of removal process using a remediating-bacterium isolated from marine coastal sediment. *RSC Adv.* 5 (46), 36665–36672.

[14] Horemans, B., Breugelmans, P., Saeys, W., Springael, D. (2017). Soil-bacterium compatibility model as a decision-making tool for soil bioremediation. *Environ. Sci. Technol.* 51 (3), 1605–1615.

[15] Tien, C. J., Huang, H. J., Chen, C. S. (2017). Accessing the carbofuran degradation ability of cultures from natural river biofilms in different environments. *Clean - Soil, Air, Water* 45 (5), 1600380.

[16] Rodríguez-Rodríguez, C. E., Madrigal-León, K., Masís-Mora, M., Pérez-Villanueva, M., Chin-Pampillo, J. S. (2017). Removal of carbamates and detoxification potential in a biomixture: fungal bioaugmentation versus traditional use. *Ecotoxicol. Environ. Saf.* 135, 252–258.

[17] Yang, L., Chen, S., Hu, M., Hao, W., Geng, P., Zhang, Y. (2011). Biodegradation of carbofuran by Pichia anomala strain HQ-C-01 and its application for bioremediation of contaminated soils. *Biology and fertility of soils*, 47, 917-923.

[18] Slaoui, M., Ouhssine, M., Berny, E. H., Elyachioui, M. (2007). Biodegradation of the carbofuran by a fungus isolated from treated soil. *Afr. J. Biotechnol.* 6 (4), 419–423.

[19] Seo, J., Jeon, J., Kim, S. D., Kang, S., Han, J., Hur, H. G. (2007). Fungal biodegradation of carbofuran and carbofuran phenol by the fungus Mucor ramannianus: identification of metabolites. *Water Sci. Technol.* 55(1-2), 163–167.

[20] Kaur, P., Balomajumder, C. (2019). Simultaneous biodegradation of mixture of carbamates by newly isolated Ascochyta sp. CBS 237.37. *Ecotoxicol. Environ. Saf.* 169, 590–599.

[21] Kaur, P., Balomajumder, C. (2020). Bioremediation process optimisation and effective reclamation of mixed carbamate-contaminated soil by newly isolated Acremonium sp., *Chemosphere,* 249, 125982.

[22] Qing, Z., Yang, L., Yu-Huan, L. (2006). Purification and characterisation of a novel Carbaryl hydrolase from Aspergillus niger PY168. *FEMS Microbiol. Lett.* 228, 39–44.

[23] Nyakundi, W. O., Magoma, G., Ochora, J., Nyende, A. B. (2011). Biodegradation of Diazinon and Methomyl Pesticides by White Rot Fungi from Selected Horticultural Farms in Rift Valley and Central Kenya. *Journal of Applied Technology in Environmental Sanitation*, 1 (2), 107-124.

[24] Tang, W. (2018). Research Progress of Microbial Degradation of Organophosphorus Pesticides. *Prog. Appl. Microbiol.* 1, 29–35.

[25] Chen, S., Hu, Q., Hu, M., Luo, J., Weng, Q., Lai, K. (2011). Isolation and Characterisation of a Fungus Able to Degrade Pyrethroids and 3-Phenoxybenzaldehyde. *Bioresour. Technol.* 102, 8110–8116.

[26] Huang, Y., Xiao, L., Li, F., Xiao, M., Lin, D., Long, X., and Wu, Z. (2018). Microbial Degradation of Pesticide Residues and an Emphasis on the

Degradation of Cypermethrin and 3-phenoxy Benzoic Acid: A Review, *Molecules*, 23, 2313.

[27] Kafilzadeh, F., Ebrahimnezhad, M., Tahery, Y. (2015). Isolation and Identification of Endosulfan-Degrading Bacteria and Evaluation of Their Bioremediation in Kor River, Iran. *Osong Public Health Res. Perspect.* 6, 39–46.

[28] Jayabarath, J., Musfira, S. A., Giridhar, R., Arulmurugan, R. (2010). Biodegradation of Carbofuran Pesticide by Saline Soil Actinomycetes. *Int. J. Biotechnol. Biochem.* 6, 187–193.

[29] Elgueta, S., Santos, C., Lima, N., Diez, M. C. (2016). Immobilisation of The White-Rot Fungus Anthracophyllum Discolor to Degrade the Herbicide Atrazine. *AMB Express*, 6, 104.

[30] Kabra, A. N., Ji, M. K., Choi, J., Kim, J. R., Govindwar, S. P., Jeon, B. H. (2014). Toxicity of Atrazine and Its Bioaccumulation and Biodegradation in A Green Microalga, Chlamydomonas Mexicana. *Environ. Sci. Pollutr.* 21, 12270–12278.

[31] Li, Z., Wang, X., Ni, Z., Bao, J. Huiwen Zhang, (2020). In-situ Remediation of Carbofuran-Contaminated Soil by Immobilized White-Rot Fungi, *Pol. J. Environ. Stud.* 29(1), 1237-1243.

[32] Castillo, MdP., Von Wiron-Lehr, S., Scheunert, I., Torstensson, L. (2001). Degradation of isoproturon by the white rot fungus Phanerochaete chrysosporium. *Biol. Fertil. Soils*, 33, 521-528.

[33] Bhalerao, T. S, Puranik, P. R. (2007). Biodegradation of organochlorine pesticide, endosulfan, by a fungal soil isolate, Aspergillus niger. *Int. Biodeterior. Biodegradation,* 59(4), 315-321.

[34] Kamaei I, Takagi K, Kondo R (2011). Degradation of endosulfan and endosulfan sulphate by white-rot fungus Trametes hirsuta. *J. Wood Sci.* 57:317.

[35] Shimizu, H. (2002). Metabolic engineering-integrating methodologies of molecular breeding and bioprocess systems engineering. *J. Biosci. Bioeng.* 94, 563-573.

[36] Urlacher, V. B., Lutz-Wahl, S., Schmid, R. D. (2004). Microbial P450 enzymes in biotechnology. *Appl. Microbiol. Biotechnol.* 64, 317-325.

[37] Alzahrani, A. M. (2009). Insects cytochrome P450 enzymes: Evolution, functions and methods of analysis. *Glob. J. Mol. Sci.* 4(2), 167-179.

[38] Zhongli, C., Shunpeng, L., Guoping, F. (2001). Isolation of methyl parathion-degrading strain m6 and cloning of the methyl parathion hydrolase gene. *Appl. Environ. Microbiol.* 67(10), 4922-4925.

[39] Pieper, D. H., Reineke, W. (2000). Engineering bacteria for bioremediation. *Curr. Opin. Biotechnol.* 11, 262-70.

[40] Shen, Y. J., Lu, P., Mei, H., Yu, H. J., Hong, Q., Li, S. P. (2009). Isolation of a methyl parathion-degrading strain Stenotrophomonas sp. SMSP-1 and cloning of the ophc2 gene. *Biodegradation* 21(5), 785-792.

[41] Eapen, S., Singh, S., D'Souza, S. F. (2007). Advances in development of transgenic plants for remediation of xenobiotic pollutants. *Biotechnol. Adv.* 25, 442-451.

[42] Cases, I., de Lorenzo, V. (2005). Promoters in the environment: Transcriptional regulation in its natural context. *Nat. Rev. Microbiol.* 3, 105-118.

[43] Chougale, V. V., Deshmukh, A. M. (2007). Biodegradation of carbofuran pesticide by saline soil actinomycetes. *Asian J. Microbiol. Biotechnol. Environ. Sci.* 9(4), 1057-1061.

[44] Wu, N. F., Deng, M. J., Liang, G. Y., Chu, X. Y., Yao, B., Fan, Y. L. (2004). Cloning and expression of ophc2, a new organophosphorus hydrolase gene. *Chin. Sci. Bull.* 49, 1245-1249.

[45] Shi, H., Pei, L., Gu, S., Zhu, S., Wang, Y., Zhang, Y., Li, B. (2012). Glutathione S-transferase (GST) genes in the red flour beetle, Triboliumcastaneum, and comparative analysis with five additional insects. *Genomics*, 100(5), 327-335.

[46] Fu, G. P., Cui, Z., Huang, T., Li, S. P. (2004). Expression, purification, and characterisation of a novel methyl parathion hydrolase. *Protein Expr. Purif.* 36(2), 170-176.

[47] Hasan, H. A. (1999). Fungal utilisation of organophosphate pesticides and their degradation by Aspergillus flavus and A. sydowii in soil. *Folia Microbial (Praha),* 44(1), 77-84.

[48] Soriano, J. M., Jiménez, B., Font, G., Moltó, J. C. (2001). Analysis of Carbamate Pesticides and Their Metabolites in Water by Solid Phase Extraction and Liquid Chromatography: A Review. *Crit. Rev. Anal. Chem.* 31(1), 19-52.

[49] Mir-Tutusaus, J. A., Masís-Mora, M., Corcellas, C., Eljarrat, E., Barceló, D., Sarrà, M., Caminal, G., Vicent, T., Rodríguez-Rodríguez, C. E. (2014). Degradation of selected agrochemicals by the white rot fungus Trametes versicolor. *The Science of the total environment*, 500-501, 235–242.

[50] Kaufman, D. D., Kearney, P. C. (1965). Microbial degradation of isopropyl-N -(3-chlorophenyl carbamate and 2-chloroethyl- N -(3-chlorophenyl carbamate. *Appl. Microbiol.* 13(3), 443-446.

[51] Kaufman, D. D., Blake, J. (1973). Microbial degradation of several acetamide, acrylamide, carbamate, toluidine, and urea pesticides. *Soil Biol. Biochem.* 5, 297-308.

[52] Li, F., Di, L., Liu, Y., Xiao, Q., Zhang, X., Ma, F., & Yu, H. (2019). Carbaryl biodegradation by Xylaria sp. BNL1 and its metabolic pathway. *Ecotoxicology and environmental safety*, 167, 331–337.

Chapter 5

The Role of Probiotics in the Detoxification of Carbamates

Yalçın Dicle*

Department of Nutrition and Dietetics, Muş Alparslan University, Muş, Turkey

Introduction

Pesticides

The rapidly increasing world population also creates an increasing need for food. Enough products cannot be supplied for this uncontrolled increase. Eliminating these problems is possible by producing high-quality, efficient, and low-cost foods [1]. However, there is a production loss of approximately 35% due to the pests in the production of some herbal products and the diseases they cause. If we do not struggle with the loss of production, this rate can double. Food deficiency can cause famine, involuntary migration, and wars [2]. For his reason, agricultural producers use various methods to minimize agricultural loss and increase the durability of these products. One of these methods is chemical control used to kill pests that live in the environment of humans, animals, and plants. These pests are parasites that carry several diseases, insects harmful to agriculture and plants, weeds and fungi, and flying and walking creatures such as flies, lice, fleas, ticks, scabies, and cockroaches around humans and animals, environments, and shelters.

The use of pesticides is one of the primary methods. The benefit of pesticide use is an undoubted fact. Pesticides play an important role in preventing and controlling potentially deadly vector-borne diseases,

* Corresponding Author's Email: y.dicle@alparslan.edu.tr.

In: The Science of Carbamates
Editor: Güllü Kaymak
ISBN: 978-1-68507-708-2

especially in high-risk tropical regions, such as dengue, trypanosomiasis, leishmaniasis, and chikungunya. Inorganic substances such as sodium chlorate, sulfuric acid, or naturally occurring organic compounds such as naphthalene, creosote, and petroleum oil were used as pesticides until the 1940s. [4]. However, it concerns people more than ever because of the accumulation of these pesticides in the human body through the consumed foods and their adverse effects on the environment [5]. Pesticides can be classified according to their target organism, mode of action, or chemical composition. Pesticides are classified into several groups according to their chemical composition: organochlorine, organophosphate, pyrethroid, triazine, neonicotinoid, carbamates [6].

Carbamates

Generally, they are compounds with drug-active properties synthesized through alcohol and carbamoyl chloride. Carbamates are stable crystalline compounds often used as recognition reagents of alcohols [7]. Among the pesticides, carbamates and their related compounds, which are less toxic than organophosphates and have lower persistence than organochlorines, have more usage areas [8].

In the pharmacological industry, carbamates are known as inactive derivatives of the drug. They are used especially as an anaesthetic agent in animals and an insecticide in the agricultural sector [9]. In addition, according to their biological effects, there are carbamates used as anti-cytotoxic [10], anti-microbial, anti-fungal, anti-bacterial [11] and acetylcholinesterase inhibitor [12]. Surgical drapes and dressings used in the medical sector are also made of polyurethane, a carbamate derivative.

Neostigmine, known as an acetylcholinesterase inhibitor that can be used in the treatment of myasthenia gravis disease, is the first synthetic drug containing carbamate in its structure synthesized by Aeschlimann and Reinert in 1931 [4]. Myasthenia gravis disease results in rapid fatigue of voluntary movement muscles, especially eye movements, chewing and swallowing movements, and causes muscle weakness. Neostigmine therapy, an anticholinesterase inhibitor, is recommended to eliminate the symptoms of this disease [13].

Rivastigmine, used in treating Alzheimer's disease, is a carbamate derivative approved by the FDA in 1999. Rivastigmine has central nervous system selectivity and is an irreversible acetylcholinesterase inhibitor [14].

Carbamate pesticides such as aldicarb, carbofuran, and ziram cause immunotoxic, carcinogenic, reproductive toxicity, and also impair the endocrine system and mitochondrial function. And even they cause disruption of mitochondrial functions and the endocrine system [15].

It has been reported that chlorpropamide, a carbamate pesticide, increases WBC count in specific dosing, decreases Hb and Hct levels in RBC count, increases spleen, liver weights, and hyperplasia of hematopoietic cell in rats. They showed that when methyl-carbamate pesticides were administered to male rats by oral, dermal, and inhalation routes, they caused dose-related decreases in thymus weight [16].

They reported that if carbofuran, a carbamate insecticide (15, 50, and 500 microgram/L), was applied to fish (*Carassius auratus*) for 24 and 48 hours, plasma glucose concentration increased, hepatic glycogen concentration decreased, and neurotoxic effects occurred in both periods (Bretaud et al., 2002) [17]. The widespread use of carbamate compounds in pest control has increased over the last two decades compared to the use of organochlorines or organophosphates [18-19].

Therefore, the emerging carbamate residues pose a threat, especially to all living creatures. The degradation/detoxification of these emerging pesticides is necessary through new technologies. There are conventional methods used to remove carbamate contaminations, but these methods are not widely accepted as they are expensive and cause post-waste generation and contamination problems [20]. For all these reasons, new microbial technologies are of great interest, instead of traditional methods, for the recovery of pesticide-contaminated sites [21]. This chapter clarifies that many bacterial species that can degrade carbamate pesticides have been isolated from the soil worldwide [22].

Carbamate-Degradable Microorganisms

The microbial degradation method is efficient mechanical remediation that can reduce and detoxify more complex organic pollutants to simpler and smaller molecules [22-24]. Since this method is performed especially in the form of on-site treatment (in-situ), it is cleaner, safer, and less destructive to the environment than traditional methods such as remediation methods of digging up [25]. Microbial methods often reduce pollutants such as carbamate to water and carbon dioxide or less toxic forms. Thus, it prevents the disposal of contaminated materials [26-27] (Figure 1).

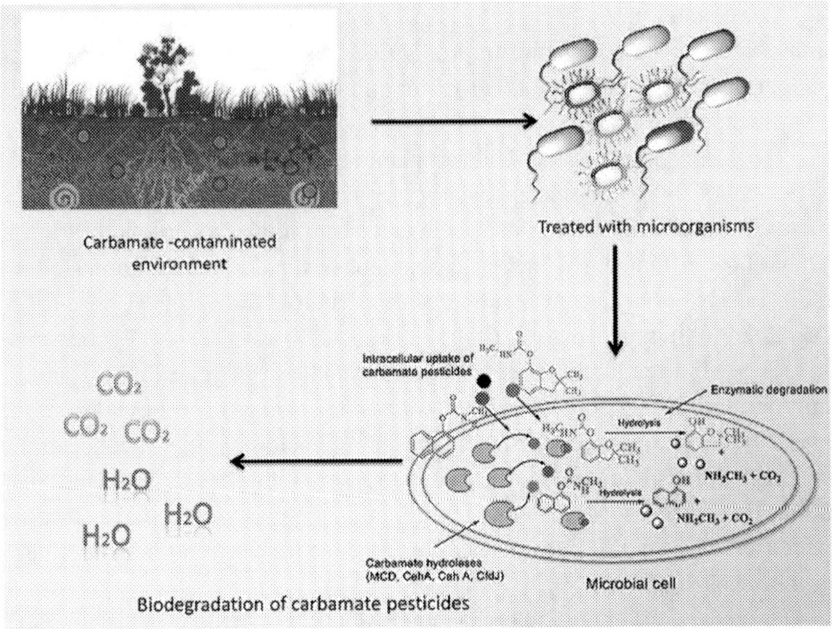

Figure 1. Microbiological degradation routes of carbamates [28].

Environmental degradation and detoxification of carbamate pesticides occur in two ways, biotic and abiotic. Processes such as adsorption, evaporation, photodegradation and alkaline hydrolysis are known as abiotic processes [29-31]. Biotic processes, which is another way, is known as the process that includes microbial degradation and completely eliminates pesticides by bacterial and fungal species. Especially to better understand the microbial degradation processes of carbamates, in many studies on soil samples exposed to pesticide contamination, Achromobacter, Aminobacter, Arthrobacter, Aspergillus, Bacillus, Burkholderia, Cupriavidus, Enterobacter, Microbacterium, Micrococcus, Mucor, Nocardioides, Novosphingobudium, Microbial species such as Serratia, Sphingbium, Sphingomonas, Stenotrophomonas, Trametes, Trichoderma have been reported to be involved in the transformation or degradation of carbamate pesticides [20, 32-52].

There are aldicarb, carbaryl, carbendazim, carbofuran, fenobucarb, methomyl, oxamyl, phenmedipham, and propoxur among the samples of some commonly used carbamate pesticides [53] (Table 1).

Table 1. Various carbamate pesticides and microbes involved in their degradation [53]

Pesticide, [IUPAC name]	Applied as	Organisms involved in degradation or biotransformation
Carbofuran [(2,2-Dimethyl-3H-1-benzo furan-7-yl)]	Insecticide	*Sphingobium* *Pichia*
Carbaryl [Naphthalen-1-yl N-methylcarbamate]	Insecticide Acaricide Nematicide	*Pseudomonas* *Micrococcus* *Aminobacter* *Trichoderma* *Aspergillus*
Propoxur [(2-Propan-2-yloxyphenyl) N-methylcarbamate]	Insecticide Acaricide Molluscicide	*Pseudaminobacter* *Nocardioides* *Pseudomonas* *Corynebacterium* *Staphylococcus* *Bacillus* *Aeromonas*
Carbendazim, [Methyl N-(1H-benzimidazol-2-yl)carbamate]	Fungicide	*Rhodococcus* *Sphingomonas* *Nocardioides* *Trichoderma*
Phenmedipham [3(methoxycarbonylamino)phenyl] N-(3-methylphenyl)carbamate	Herbicide	*Arthrobacter* *Pseudomonas* *Ochtrobactrum* *Trichoderma*
Fenobucarb [(2-Butan-2-ylphenyl) N methylcarbamate]	Insecticide	*Sphingobium* *Pichia*
Oxamyl [Methyl 2-(dimethylamino)-N-(methylcarbamoyloxy)-2- oxoethanimido thioate]	Insecticide Acaricide Nematicide	*Pseudomonas* *Micrococcus* *Aminobacter* *Trichoderma* *Aspergillus*
Methomyl, [Methyl N-(methylcarbamoyloxy) ethanimido thioate]	Insecticide Acaricide Nematicide	*Paracoccus* *Pseudomonas* *Aminobacter* *Flavobacterium* *Alcaligenes* *Bacillus* *Serratia* *Novosphingobium* *Trametes*
Aldicarb [(E)-(2-Methyl-2-methylsulfanylpropylidene)amino N-methylcarbamate]	Insecticide Acaricide Nematicide	*Enterobacter* *Stenotrophomonas* *Methylosinus*

Probiotic Microorganisms as Detoxification Tools in the Degradation and Transformation of Carbamates

The probiotics are beneficial live microorganisms that have positive effects on the gastrointestinal microflora of the host when taken into the body in specific numbers [54]. Among these microorganisms, the most commonly used bacteria are selected strains of *Lactobacillus* and *Bifidobacterium* species. Functional effects of probiotic bacteria in terms of health care are lactose intolerance, reducing the serum cholesterol level, controlling the gastrointestinal infections, enhancing the immune system, showing anti-microbial, antimutagenic, antiallergic, and anticarcinogenic activities, increasing the biological value of nutrients, producing vitamins, and increasing the utilization of minerals and trace elements. [55-56]. In addition to the positive effects of these bacteria on health, their ability to degrade the food components non-used by humans and have toxic effects into smaller molecules with no toxic effect or can be digested by humans; in other words, their biological detoxification effect draws attention [57].

There exist some physical, chemical, and biological risks threatening safe production, transportation, storage, distribution, and consumption of foods. These risks cause significant economic losses for countries as well as human health [58]. Therefore, physical, chemical, and biological detoxification processes are applied to prevent and/or reduce the possible adverse effects of risks. Detoxification is the elimination or deactivation process of harmful components such as drugs, mutagens, carcinogens, etc., from the human body and food [59].

Paralytic shellfish toxins (PSTs) are produced by blue-green algae and fiery algae. These toxins are chemically divided into three main groups as carbamate toxins (saxitoxin, neosaxitoxin and gonatoxin), sulfocarbamoyl toxins, and decarbamoyl toxins (C-toxin) [60-61]. In the toxin binding studies of probiotics, the binding capacities of saxitoxin, neosaxitoxin, gonatoxin and C-toxin were demonstrated by using living and non-living forms of Lb. rhamnosus GG and LC-705 strains. In conclusion, it was determined that while neosaxitoxin and saxitoxin were bound by 77-97.2%, goniatoxin and C-toxin were bound by 33.3-49.7%. While there was no significant difference between the capacity to bind toxins between living and non-living forms, this shows that toxins are removed by binding capacity rather than metabolism [62].

Fermentation is an ancient food preparation process reported to be an effective method in reducing pesticide levels. In the studies, the degradation of pirimiphos-methyl by *L Plantarum* during wheat fermentation was evaluated. The results showed that 81% of the pesticide was degraded by L Plantarum bacteria species, and this pesticide did not affect bacterial growth and fermentation activity [63]. In a similar study, the degradation of bifenthrin, a pyrethroid insecticide, has been observed in ground wheat during fermentation by *L. Plantarum*. It is concluded that bifenthrin is metabolized by hydrolysis of the carboxyl ester linkage [64].

It is stated that some lactic acid bacteria (LAB) are belonging to *Lactobacillu*s and *Leuconostoc* genera can metabolize insecticides and can also be used as a carbon and energy source [65].

Various species in the genera of *Akkermansia, Bacillus, Bacteroides, Bifidobacterium, Enterococcus, Lactobacillus, Lactococcus, Pediococcus, Propionibacterium, Saccharomyces, Streptococcus* and *Streptococcus* were considered as probiotics, and the most common ones are specified as *Bifidobacterium animalis, Lactobacillus acidophilus, Lactobacillus bulgaricus, Lactobacillus casei, Lactobacillus johnsoniin, Lactobacillus lactis, Lactobacillus plantarum, Lactobacillus reuteri* and *Lactobacillus salivarius* species [66].

It has been reported that many enzymes such as phosphatases, phosphotriesterases, carboxylesterases and organophosphate hydrolases are involved in the degradation of organophosphate pesticides by hydrolysis of phosphoric acid esters. The COP linkage of a wide variety of phosphate esters of organophosphate pesticides can be hydrolyzed and disrupted by both acid and alkaline phosphatases [65, 67]. Probiotics have pesticide degrading genes that encode enzymes with degrading functions such as organophosphorus acid anhydrases, methyl parathion hydrolases, and organophosphorus hydrolases. It has been reported that gene expression increases in probiotics, especially in the presence of pesticides in the growth medium [68].

Studies proved that pesticides could interact with the gut microbiota, alter its composition and metabolites, including short-chain fatty acids, bile acids, and trimethylamine, affect the receptor sites of different tissues and organs, and destroy intestinal mucosa and cells by leading to pathological changes [69]. For instance, chlorpyrifos has been reported to decrease the number of beneficial bacteria (*Bifidobacterium* and *Lactobacillus*) and increase the numbers of *Enterococcus* and *Bacteroide*s [70].

Conclusion

There is an increasing interest in using probiotic bacteria for food safety and as nutritional supplements due to their positive effects on health and protective properties. Removal of chemical and microbial toxins using probiotic bacteria is introduced as a new, promising biological method that can be an alternative to conventional decontamination methods. Studies demonstrate that probiotic bacteria can prevent the development of harmful microorganisms, reduce toxic metabolites formed by these harmful microorganisms, and can be used as a protective agent for food safety.

References

[1] Vural, N. T. (2005). *Toksikoloji.* Ankara Üniversitesi Eczacılık Fakültesi Yayınları. Ankara. 342-373.

[2] Birişik, N., Özdem, A., Karahan, A., Sezgen, M., Ertürk, S., Alkan, M. & Bayram, Y. (2018). *Teoriden pratiğe kimyasal mücadele.*1. Baskı. Gıda Tarım ve Hayvancılık Genel Müdürlüğü.

[3] Toros, S., Maden, S., & Sözeri, S. (1999). *Tarımsal savaş yöntem ve ilaçları.* Ankara, AÜ Ziraat Fakültesi Yayınları, 27-42.

[4] Unsworth, J. (2010). *History of Pesticide Use. International Union of Pure and Applied Chemistry (IUPAC).* Available online at http://agrochemicals.iupac.org/index.php?option=com_sobi2&sobi2Task=sobi2Details&catid=3&sobi2Id=31 (Accessed September, 2021).

[5] Ogut, S. *Pestisitlerin Olumsuz Sağlık ve Çevre Etkileri.* (2009). Available online at https://akademik.adu.edu.tr/aum/sesam/default.asp?idx=323131 (Accessed September, 2021).

[6] Arias-Estévez, M., López-Periago, E., Martínez-Carballo, E., Simal-Gándara, J., Mejuto, J. C., & García-Río, L. (2008). The mobility and degradation of pesticides in soils and the pollution of groundwater resources. *Agriculture, ecosystems & environment*, 123(4), 247-260.

[7] Solomons, T. W. G., Fryhle, C. B., (2002). *Organik Kimya.* 7. Baskıdan Çeviri. Literatür Yayınları, İstanbul, 783-785.

[8] Hernández, A. F., Parrón, T., Tsatsakis, A. M., Requena, M., Alarcón, R., & López-Guarnido, O. (2013). Toxic effects of pesticide mixtures at a molecular level: their relevance to human health. *Toxicology*, 307, 136-145.

[9] Perveen, S., Yasmin, A., Khan, K. M., (2010), Effect of successive increase in alcohol chains on reaction with isocyanates and isothiocyanates, *Natural Product Research.* 24, 18-23.

[10] Liu, J. F., Sang, C. Y., Xu, X. H., Zhang, L. L., Yang, X., Hui, L., Zhang, J. B., Chen, S. W., (2013), Synthesis and cytotoxic activity on human cancer cells of carbamate derivatives of 4b-(1,2,3- triazol-1-yl)podophyllotoxin, European *Journal of Medicinal Chemistry*. 64, 621-628.

[11] Krátký, M., Volková, M., Novotná, E., Trejtnar, F., Stolaříková, J., & Vinšová, J. (2014). Synthesis and biological activity of new salicylanilide N, N-disubstituted carbamates and thiocarbamates. *Bioorganic & medicinal chemistry*, *22*(15), 4073-4082.

[12] Mustazza, C., Borioni, A., Giudice, M. R. D., Gatta, F., Ferretti, R., Meneguz, A., Volpe, M. T., Lorenzini, P., (2002), Synthesis and cholinesterase activity of phenylcarbamates related to Rivastigmine, a therapeutic agent for Alzheimer's disease, *Eur. J. Med. Chem.* 37, 91-109.

[13] Conti-Fine, B. M., Milani, M., & Kaminski, H. J. (2006). Myasthenia gravis: past, present, and future. *The Journal of clinical investigation*, 116(11), 2843-2854.

[14] Rösler, M., Bayer, T., Anand, R., Cicin-Sain, A., Gauthier, S., Agid, Y., & Gharabawi, M. (1999). Efficacy and safety of rivastigmine in patients with Alzheimer's disease: international randomised controlled trialCommentary: Another piece of the Alzheimer's jigsaw. *Bmj*, 318(7184), 633-640.

[15] Nicolopoulou-Stamati, P., Maipas, S., Kotampasi, C., Stamatis, P., & Hens, L. (2016). Chemical pesticides and human health: the urgent need for a new concept in agriculture. *Frontiers in public health*, 4,148.

[16] Ladics, G. S., Smith, C., Heaps, K., & Loveless, S. E. (1994). Evaluation of the humoral immune response of CD rats following a 2-week exposure to the pesticide carbaryl by the oral, dermal, or inhalation routes. *Journal of Toxicology and Environmental Health, Part A Current Issues*, 42(2), 143-156.

[17] Bretaud, S., Saglio, P., Saligaut, C., & Auperin, B. (2002). Biochemical and behavioral effects of carbofuran in goldfish (*Carassius auratus*). *Environmental Toxicology and Chemistry: An International Journal*, 21(1), 175-181.

[18] Silberman, J., & Taylor, A. (2018). *Carbamate toxicity*.

[19] Carazo-Rojas, E., Pérez-Rojas, G., Pérez-Villanueva, M., Chinchilla-Soto, C., Chin-Pampillo, J. S., Aguilar-Mora, P., ... & Vryzas, Z. (2018). Pesticide monitoring and ecotoxicological risk assessment in surface water bodies and sediments of a tropical agro-ecosystem. *Environmental pollution*, 241, 800-809.

[20] Mustapha, M. U., Halimoon, N., Johar, W. L. W., & Abd Shukor, M. Y. (2019). An Overview on Biodegradation of Carbamate Pesticides by Soil Bacteria. *Pertanika Journal of Science & Technology*, 27(2019), 547-563.

[21] Pang, S., Lin, Z., Zhang, W., Mishra, S., Bhatt, P., & Chen, S. (2020). Insights into the microbial degradation and biochemical mechanisms of neonicotinoids. *Frontiers in microbiology*, 11, 868.

[22] Desaint, S., Hartmann, A., Parekh, N. R., & Fournier, J. C. (2000). Genetic diversity of carbofuran-degrading soil bacteria. *FEMS Microbiology Ecology*, 34(2), 173-180.

[23] Zhan, H., Wang, H., Liao, L., Feng, Y., Fan, X., Zhang, L., & Chen, S. (2018). Kinetics and novel degradation pathway of permethrin in Acinetobacter baumannii ZH-14. *Frontiers in microbiology*, 9, 98.

[24] Mishra, S., Zhang, W., Lin, Z., Pang, S., Huang, Y., Bhatt, P., & Chen, S. (2020). Carbofuran toxicity and its microbial degradation in contaminated environments. *Chemosphere*, 259, 127419.

[25] Sharma, J. (2019). Advantages and limitations of in situ methods of bioremediation. *Recent Adv Biol Med*, 5(2019), 10941.

[26] Azubuike, C. C., Chikere, C. B., & Okpokwasili, G. C. (2016). Bioremediation techniques-classification based on site of application: principles, advantages, limitations and prospects. *World Journal of Microbiology and Biotechnology*, 32(11), 1-18.

[27] Parte, S. G., Mohekar, A. D., & Kharat, A. S. (2017). Microbial degradation of pesticide: a review. *African journal of microbiology research*, 11(24), 992-1012.

[28] Mishra, S., Pang, S., Zhang, W., Lin, Z., Bhatt, P., & Chen, S. (2021). Insights into the microbial degradation and biochemical mechanisms of carbamates. *Chemosphere*, 130500.

[29] Faust, S. D., & Gomaa, H. M. (1972). Chemical hydrolysis of some organic phosphorus and carbamate pesticides in aquatic environments. *Environmental letters*, 3(3), 171-201.

[30] Raut-Jadhav, S., Pinjari, D. V., Saini, D. R., Sonawane, S. H., & Pandit, A. B. (2016). Intensification of degradation of methomyl (carbamate group pesticide) by using the combination of ultrasonic cavitation and process intensifying additives. *Ultrasonics sonochemistry*, 31, 135-142.

[31] Tomašević, A., Mijin, D., Marinković, A., Cvijetić, I., & Gašić, S. (2019). Photocatalytic degradation of carbamate insecticides: effect of different parameters. *Pesticidi i fitomedicina*, 34(3-4), 193-200.

[32] Kim, I. S., Ryu, J. Y., Hur, H. G., Gu, M. B., Kim, S. D., & Shim, J. H. (2004). Sphingomonas sp. strain SB5 degrades carbofuran to a new metabolite by hydrolysis at the furanyl ring. *Journal of agricultural and food chemistry*, 52(8), 2309-2314.

[33] Yan, X., Jin, W., Wu, G., Jiang, W., Yang, Z., Ji, J., & Hong, Q. (2018). Hydrolase CehA and monooxygenase CfdC are responsible for carbofuran

degradation in Sphingomonas sp. strain CDS-1. *Applied and environmental microbiology*, 84(16), e00805-18.

[34] Singh, R., Trivedi, V. D., & Phale, P. S. (2013). Metabolic regulation and chromosomal localization of carbaryl degradation pathway in Pseudomonas sp. strains C4, C5 and C6. *Archives of microbiology*, 195(8), 521-535.

[35] Naqvi, T. A., Armughan, A., Ahmed, N., & Ahmed, S. (2013). Biodegradation of carbamates by Pseudomonas aeruginosa. *Minerva Biotecnol*, 25(4), 207-211.

[36] Zhu, S., Qiu, J., Wang, H., Wang, X., Jin, W., Zhang, Y., & Hong, Q. (2018). Cloning and expression of the carbaryl hydrolase gene mcbA and the identification of a key amino acid necessary for carbaryl hydrolysis. *Journal of hazardous materials*, 344, 1126-1135.

[37] Mohamed, M. S. (2009). Degradation of methomyl by the novel bacterial strain Stenotrophomonas maltophilia M1. *Electronic Journal of Biotechnology*, 12(4), 1-6.

[38] Fareed, A., Zaffar, H., Rashid, A., Maroof Shah, M., & Naqvi, T. A. (2017). Biodegradation of N-methylated carbamates by free and immobilized cells of newly isolated strain *Enterobacter cloacae* strain TA7. *Bioremediation Journal*, 21(3-4), 119-127.

[39] Ekram, M. A. E., Sarker, I., Rahi, M. S., Rahman, M. A., Saha, A. K., & Reza, M. A. (2020). Efficacy of soil-borne Enterobacter sp. for carbofuran degradation: HPLC quantitation of degradation rate. *Journal of basic microbiology*, 60(5), 390-399.

[40] Naqvi, T., Cheesman, M. J., Williams, M. R., Campbell, P. M., Ahmed, S., Russell, R. J., & Oakeshott, J. G. (2009). Heterologous expression of the methyl carbamate-degrading hydrolase MCD. *Journal of biotechnology*, 144(2), 89-95.

[41] Lawrence, K. S., Feng, Y., Lawrence, G. W., Burmester, C. H., & Norwood, S. H. (2005). Accelerated degradation of aldicarb and its metabolites in cotton field soils. *Journal of nematology*, 37(2), 190-197.

[42] Hayatsu, M., Hirano, M., & Nagata, T. (1999). Involvement of two plasmids in the degradation of carbaryl by Arthrobacter sp. strain RC100. *Applied and Environmental Microbiology*, 65(3), 1015-1019.

[43] Hayatsu, M., Mizutani, A., Hashimoto, M., Sato, K., & Hayano, K. (2001). Purification and characterization of carbaryl hydrolase from Arthrobacter sp. RC100. *FEMS microbiology letters*, 201(1), 99-103.

[44] Osborn, R. K., Haydock, P. P. J., & Edwards, S. G. (2010). Isolation and identification of oxamyl-degrading bacteria from UK agricultural soils. *Soil Biology and Biochemistry*, 42(6), 998-1000.

[45] Zhang, C., Yang, Z., Jin, W., Wang, X., Zhang, Y., Zhu, S. & Hong, Q. (2017). Degradation of methomyl by the combination of Aminobacter sp.

MDW-2 and Afipia sp. MDW-3. *Letters in applied microbiology*, 64(4), 289-296.

[46] Nguyen, T. P. O., De Mot, R., & Springael, D. (2015). Draft genome sequence of the carbofuran-mineralizing Novosphingobium sp. strain KN65. 2. *Genome announcements*, 3(4), e00764-15.

[47] Roy, T., & Das, N. (2017). Isolation, characterization, and identification of two methomyl-degrading bacteria from a pesticide-treated crop field in west Bengal, India. *Microbiology*, 86(6), 753-764.

[48] Mohamed, E. A. (2017). Oxamyl Utilization by Micrococcus luteus OX, Isolated from Egyptian Soil. *International Journal of Applied Environmental Sciences*, 12(5), 999-1008.

[49] Yang, C., Xu, X., Liu, Y., Jiang, H., Wu, Y., Xu, P., & Liu, R. (2017). Simultaneous hydrolysis of carbaryl and chlorpyrifos by Stenotrophomonas sp. strain YC-1 with surface-displayed carbaryl hydrolase. *Scientific reports*, 7(1), 1-8.

[50] Gupta, J., Rathour, R., Singh, R., & Thakur, I. S. (2019). Production and characterization of extracellular polymeric substances (EPS) generated by a carbofuran degrading strain Cupriavidus sp. ISTL7. *Bioresource technology*, 282, 417-424.

[51] Jiang, W., Gao, Q., Zhang, L., Wang, H., Zhang, M., Liu, X., ... & Hong, Q. (2020). Identification of the key amino acid sites of the carbofuran hydrolase CehA from a newly isolated carbofuran-degrading strain Sphingbium sp. CFD-1. *Ecotoxicology and environmental safety*, 189, 109938.

[52] Seo, J. S., Keum, Y. S., & Li, Q. X. (2013). Metabolomic and proteomic insights into carbaryl catabolism by Burkholderia sp. C3 and degradation of ten N-methylcarbamates. *Biodegradation*, 24(6), 795-811.

[53] Malhotra, H., Kaur, S., & Phale, P. S. (2021). Conserved metabolic and evolutionary themes in microbial degradation of carbamate pesticides. *Frontiers in Microbiology*, 12;1863.

[54] Barros, C. P., Guimarães, J. T., Esmerino, E. A., Duarte, M. C. K., Silva, M. C., Silva, R., ... & Cruz, A. G. (2020). Paraprobiotics and postbiotics: concepts and potential applications in dairy products. *Current Opinion in Food Science*, 32, 1-8.

[55] Delikanli, B., & Özcan, T. (2014). Probiyotik içeren yenilebilir filmler ve kaplamalar. *Uludağ Üniversitesi Ziraat Fakültesi Dergisi*, 28(2), 59-70.

[56] Soccol, C. R., Prado, M. R., Garcia, L. M., Rodrigues, C., Medeiros, A. B. P., & Soccol, V. T. (2014). Current developments in probiotics. *J. Microb. Biochem. Technol*, 7, 11-20.

[57] Zoghi, A., Khosravi-Darani, K., & Sohrabvandi, S. (2014). Surface binding of toxins and heavy metals by probiotics. *Mini reviews in medicinal chemistry*, 14(1), 84-98.

[58] Erkmen, O. (2010). Gıda kaynaklı tehlikeler ve güvenli gıda üretimi. *Çocuk Sağlığı ve Hastalıkları Dergisi*, 53(3), 220-235.

[59] Phelps, K., & Hassed, C. (2012). *Substance Misuse: General Practice: The Integrative Approach Series*. Elsevier Health Sciences. 54 s.

[60] Bartram, J. and I. Chorus. (2002). *Toxic Cyanobacteria in water: A Guide To Their Public Health Consequences, Monitoring And Management.* Taylor & Francis: Oxford, UK.

[61] Kodama, M. (2010). *Paralytic shellfish poisoning toxins: biochemistry and origin.* Terrapub.

[62] Vasama, M., Kumar, H., Salminen, S., & Haskard, C. A. (2014). Removal of paralytic shellfish toxins by probiotic lactic acid bacteria. *Toxins*, 6(7), 2127-2136.

[63] Đorđević, T. M., Šiler-Marinković, S. S., Đurović-Pejčev, R. D., Dimitrijević-Branković, S. I., & Gajić Umiljendić, J. S. (2013a). Dissipation of pirimiphos-methyl during wheat fermentation by L actobacillus plantarum. *Letters in applied microbiology*, 57(5), 412-419.

[64] Đorđević, T. M., Šiler-Marinković, S. S., Đurović, R. D., Dimitrijević-Branković, S. I., & Gajić Umiljendić, J. S. (2013b). Stability of the pyrethroid pesticide bifenthrin in milled wheat during thermal processing, yeast and lactic acid fermentation, and storage. *Journal of the Science of Food and Agriculture*, 93(13), 3377-3383.

[65] Islam, S. M. A., Math, R. K., Cho, K. M., Lim, W. J., Hong, S. Y., Kim, J. M., ... & Yun, H. D. (2010). Organophosphorus hydrolase (OpdB) of Lactobacillus brevis WCP902 from kimchi is able to degrade organophosphorus pesticides. *Journal of agricultural and food chemistry*, 58(9), 5380-5386.

[66] Khaneghah, A. M., Abhari, K., Eş, I., Soares, M. B., Oliveira, R. B., Hosseini, H., ... & Sant'Ana, A. S. (2020). Interactions between probiotics and pathogenic microorganisms in hosts and foods: A review. *Trends in Food Science & Technology*, 95, 205-218.

[67] Lili, Z., Junyan, W., Hongfei, Z., Baoqing, Z., & Bolin, Z. (2018). Detoxification of cancerogenic compounds by lactic acid bacteria strains. *Critical reviews in food science and nutrition*, 58(16), 2727-2742.

[68] Sidhu, G. K., Singh, S., Kumar, V., Dhanjal, D. S., Datta, S., & Singh, J. (2019). Toxicity, monitoring and biodegradation of organophosphate pesticides: a review. *Critical reviews in environmental science and technology*, 49(13), 1135-1187.

[69] King, A. M., & Aaron, C. K. (2015). Organophosphate and carbamate poisoning. *Emergency Medicine Clinics*, 33(1), 133-151.

[70] Hlihor, R. M., Gavrilescu, M., Tavares, T., Favier, L., & Olivieri, G. (2017). Bioremediation: an overview on current practices, advances, and new perspectives in environmental pollution treatment. *BioMed research international.*

Chapter 6

Genotoxic Effects of Carbamate Insecticides

Serdal Öğüt*
Aydın Adnan Menderes University, Faculty of Health Sciences, Aydın, Turkey

Introduction

The increase in the world's population every day increases the demand for food. Therefore, there has been an increase in the use of pesticides in recent years [1-2]. It is reported that the pesticide use reached 4.5 million tons worldwide by 1990-2017 [3].

People's recognition of pesticides goes back centuries. The earliest known use is elemental sulphur, which was used by ancient Sumerians about 4500 years ago in Mesopotamia [4].

Pesticides are non-specific chemicals used to control pests. Pesticides are used to combat pests in agriculture, at home and in gardens. Every person wants to adopt a healthy lifestyle [5]. Intensive and false of pesticides threatens human health [2].

Pesticides are subject to a wide variety of classifications. There is a wide variety of classifications based on its structure, target organism, function and hazard size. However, the most widely used and best-known classification type it is a classification made according to the organisms they target. Accordingly, pesticides are divided into various classes such as herbicides, insecticides, fungicides, acaricides, nematicides, rodenticides, mollucides.

* Corresponding Author's Email: serdal.ogut@adu.edu.tr.

In: The Science of Carbamates
Editor: Güllü Kaymak
ISBN: 978-1-68507-708-2

Pesticides and Health

When pesticides are used incorrectly, they negatively affect water, air, environment and organism health. Decomposition products can harm people and nature. They can get into the air and pollute the air we breathe. Worst of all, they can harm the health of many living things. People can be adversely affected by pesticides through direct or indirect exposure [6].

Acute or chronic poisoning occurs when foods containing high doses of pesticides are ingested into the human body. Acute poisonings have a rapid and severe effect on the organism. They cause the direct death of the living thing [7]. Chronic poisoning, on the other hand, affects the organism slowly and in a long time. It shows this effect by causing various serious diseases in humans. Cancer, nervous and immune system disorders, skin problems are examples of these serious diseases [8]. In addition, pesticides can create reactive oxygen species (ROS) that cause oxidative stress and cause changes in antioxidants [9]. In addition, pesticides have many different negative effects on health. Health effects of pesticides are given in figure 1 [10].

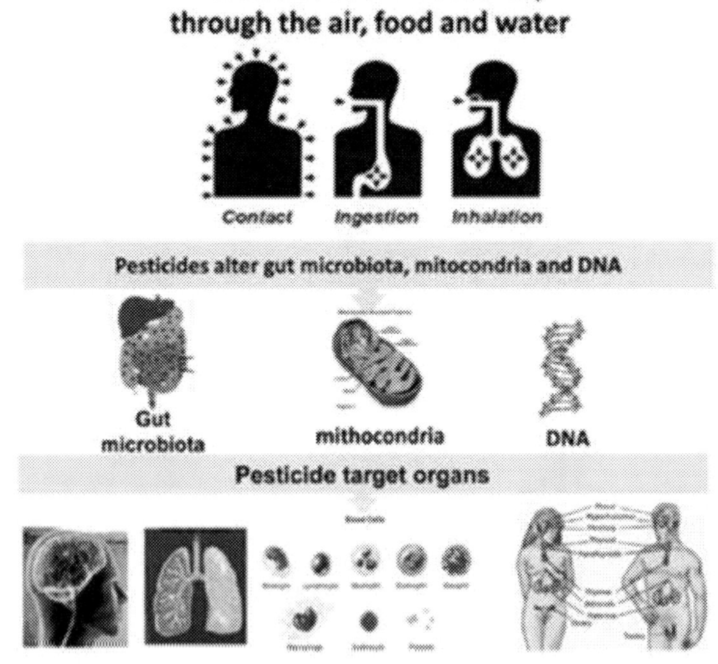

Figure 1. Health effects of pesticides.

Carbamate Insecticides

Carbamates are used in agriculture as insecticides, herbicides, and nematicides. In addition, it is also used against pests in workplaces. Their general formula is R1NHCOOR2. It is possible to divide these compounds into three as Insecticides which are ester derivatives of Carbamate, Carbamate Herbicides and Carbamate Fungicides [11]. The some carbamate insecticides and their chemical structures are given in the figure 2.

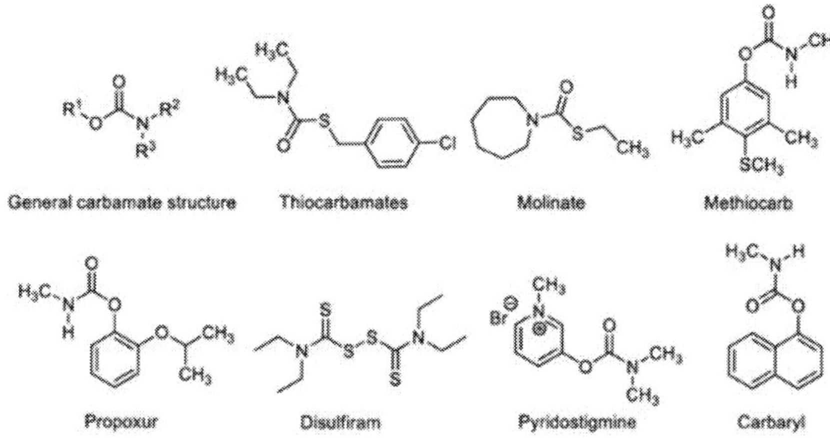

Figure 2. Some carbamate insecticides and their chemical structures [12].

Table 1. Some carbamate pesticides with their effects are given [13]

Pesticide name	Lethal dose (LD50)	Health Effects
Aldicarb	0.84 mg/kg	Nausea, Weakness, Tremors, Paralysis in respiratory system, Headache, Cancer
Bendiocarb	34-64 mg/kg	Inhibits Acetylcholinesterase enzyme activity
Carbaryl	850 mg/kg	Inhibit Cholinesterase enzyme and Carcinogenic
Carbofuran	8-14 mg/kg	Disrupt The normal functioning of hormones, decreases sperm count and sperm motility
Carbosulfan	90-250 mg/kg	Reversible inhibition of cholinesterase enzyme

Researches on the Genotoxic Effects of Carbamate Insecticides

In a study conducted on peripheral blood samples taken from healthy people, the genotoxic effects of some carbamate pesticides (aldicarb sulfon, ethiofencarb) were investigated. The Comet technique was used for the research, and DNA strand breaks and damages in the alkali labile regions were detected in the cells. In the study, it was observed that these pesticides increased DNA damage depending on the incubation time and dose [14].

Carbamate insecticide methomyl could induce genotoxic effects, including micronuclei, chromosome aberrations and sister-chromatid exchanges. Guanggang et al., (2013) investigated the genotoxic effects of methomyl in drosophila. The alkaline comet assay was used to evaluate total DNA single strand breaks (SSBs) in research. In conclusion, methomyl was a strongly genotoxic agent that induces cell DNA damage and apoptosis in vitro (Drosophila S2, HeLa and HEK293 cells) [15].

In another study, genotoxicity of the carbamate insecticide methiocarb in *Allium cepa* was investigated. As a result methiocarb administration caused an increase in the number of micronucleus (MN) and chromosomal damage [16].

Genotoxic effects of carbaryl on human umbilical vein endothelial cells (HUVECs) researched. HUVECs exposed for 24 h with low to high concentrations (10, 25, 50, 100, 250, 500, and 1000 μM) of carbaryl exhibited significant proliferation inhibition only at the higher concentrations (500 and 1000 μM) [17].

Carbaryl exhibited cytotoxicity and transcriptional changes in the neuronal developmental genes of human induced pluripotent stem cells (iPSCs) [18]. Concerning the toxicity in non-target species, carbaryl causes histopathological, cytopathological, and ultrastructural anomalies in *Oreochromis niloticus* [19].

Neuronal cell cultures of the rat were used in the study in which the effects of fenoxycarb were investigated. In this cell culture model endpoints such as viability, energy supply, glucose consumption, glutathione (GSH) levels and cytoskeleton elements were determined. Added to cultured rat cortical neurons for 1 week, fenoxycarb and propoxur considerably decreased ATP levels, mitochondrial membrane potential and glucose consumption. Besides this, fenoxycarb had an impact on neurofilaments [20].

Aydemir and Bilaloğlu (2004) researched the genotoxic effects of propamocarb. They were carried out bone-marrow micronucleus and chromosome aberration test in Swiss albino mice. Mice were injected with four different doses of propamocarb intraperitoneally; 50, 100, 200 and 400 mg/kg b.w. Propamocarb did not increase the frequency of micronucleated erythrocytes, but decreased the polychromatic/normochromatic erythrocytes ratio at all sampling intervals. Propamocarb increased only gaps in total chromosome aberrations [21].

In general, published data on carbofuran genotoxicity are scarce. There are only three studies evaluating its apoptotic potential, all of which could support our observation.

Yoon et al., (2001) and In a reserch have shown that in Chinese hamster lung cell lines, N-nitrosocarbofuran (NOCF), as the metabolite of carbofuran, efficiently activates caspase-9, caspase-8 and caspase-3, and subsequent cleavage of poly(ADP-ribose) polymerase, thus inducing apoptosis [22]. The authors detected a dose-dependent increase in cytosolic cytochrome c, suggesting that the mitochondrial pathway is primarily involved in NOCF-induced apoptosis [23].

In other study genotoxic effects of acute carbofuran intoxication in man researched. Considering the results of this study, applying genotoxicity assays in addition to the analysis of ChE activity in routine periodical monitoring of carbamate exposed populations is strongly recommended [24]. In this research, fluorescence in situ hybridization (FISH) characterization applying pancentromeric probes of genome damage in binucleated lymphocytes of a worker intoxicated by inhalation of carbofuran in the course of pesticide production [24].

Sun et al., (2010) investigated the genotoxic properties of aldicarb. In comet assay, high concentration groups of aldicarb resulted in different degrees of DNA damage of human peripheral blood lymphocytes [25].

DNA damage (*in vitro* via by Comet test) in human leukocytes was investigated in a study with bendiocarb. Leukocyte isolation was performed from blood samples taken from six healthy male individuals who did not smoke and use alcohol and were not exposed to any chemicals in the working environment, and then their leukocytes exposed to increasing doses of bendiocarb (20, 40, 80, 160 μg/ mL). In the Comet test, DNA tail percentage, length and moment were increased significantly in increasing doses of bendiocarb and DNA damage was detected ($p < 0.05$) [26]. In this research comet analysis of the effect of increasing doses of bendiocarb on the DNA of human leukocyte cells.

Conclusion

Biomonitoring of genotoxicity in living things is an important tool for determining the genetic risk that may occur as a result of exposure to pesticides. As with many pesticide groups, carbamate pesticides also show genotoxic effects in *in vitro* and *in vivo* studies. Cell culture experiments, studies on rats and humans clearly demonstrate these genotoxic effects. The most important measure to be taken at this stage is to use pesticides at recommended doses. In addition, pesticide applicators should take the necessary precautions and apply pesticides. Consumers should also be made aware of possible pesticide residue problems.

References

[1] Ögüt S, Küçüköner E, Gültekin F. 2012. The effects of pesticides on greenhouse workers and their produced products. *Toxicol. Environ. Chem.*, 94, 403–410.

[2] Varol E, Ogut S, Gultekin F. 2014. Effect of pesticide exposure on platelet indices in farm workers. *Toxicology and Industrial Health*, 30(7):630-634.

[3] *Food and Agriculture Data for Over 245 Countries FAO*. 2017. Available from: http://www.fao.org/faostat/en/#data/RP/visualize.

[4] Rao G V, Rupela O P, Rao V R, Reddy Y V. 2007. Role of biopesticides in crop protection: present status and future prospects. *Indian Journal of Plant Protection*, 35(1):1–9.

[5] Duran Ü, Öğüt S, Asgarpour H. Kunter D. 2018. Evaluation of the Health Personnel's Healthy Lifestyle Behaviors. *Journal of Adnan Menderes University Health Sciences Faculty*, 2(3); 138-147.

[6] Öğüt S. Genotoxic effects of pesticides. 2019. *Journal of Environmental Protection and Ecology* 20, No 1, 224–229.

[7] Abdollahi M, Ranjbar A, Shadnia S, Nikfar S, Rezaiee A. 2004. Pesticides and oxidative stress: a review. *Medical Science Monitor*, 10(6): RA141-RA147.

[8] Ündeğer Ü, Schlumpf M, Lichtensteiger W. 2010. Effect of the herbicide pendimethalin on rat uterine weight and gene expression and in silico receptor binding analysis. *Food and chemical toxicology*, 48(2):502-508.

[9] Öğüt S, Kucukoner E, Gultekın F, Gurbuz: N. 2015. Study of Long-term Pesticide Application amongst Agricultural Workers: Total Antioxidant Status, Total Oxidant Status and Acetylcholinesterase Activity in Blood. *Proc. Natl. Acad. Sci. India Sect. B Biol. Sci.*, 85, 155.

[10] Alleva R, Manzella N, Gaetani S, Bacchetti T, Bracci M, Ciarapica V, Monaco F, Borghi B, Amati M, Ferretti G, Tomasetti M. 2018. Mechanism underlying the effect of long-term exposure to low dose of pesticides on DNA integrity. *Environ. Toxicol.* Apr;33(4):476-487. doi: 10.1002/tox.22534. Epub 2018 Jan 23. Erratum in: *Environ. Toxicol.* 2018 Dec;33(12):1329. PMID: 29359425.

[11] Vural, N, Toksikoloji. 2005. 975-482-289-1, Ankara Üniversitesi Eczacılık Fakültesi Yayınları No:73, Ankara, 373-377.

[12] Vlcek V, Pohanka M. 2012. Carbamate insecticides in the czech republic: health and environmental impacts. *Mil. Med. Sci. Lett. (Voj. Zdrav. Listy)* 81(1), p. 2-8.

[13] Khanam S. 2019. Behavioural Changes in Broiler Chicks Exposed to Carbaryl. *American Journal of Biomedical Science and Research*, 3(3), 209-212, ISSN: 2642-1747. USA.

[14] İncedere, F E, Bazı Pestisitlerin Genotoksik Etkilerinin Araştırılması. 2009. Yüksek Lisans Tezi, MÜ Sağlık Bilimleri Enstitüsü, İstanbul, Turkiye,

[15] Guanggang X C, Diqiu L, Jianzhong Y, Jingmin G, Huifeng Z, Mingan S, Liming T. 2013. Carbamate insecticide methomyl confers cytotoxicity through DNA damage induction. *Food and Chemical Toxicology*, 53, 352-358.

[16] Tütüncü E, Yalçin E, Acar A, Yapar K, Çavuşoğlu K. 2019. Investigation of the Toxic Effects of a Carbamate Insecticide Methiocarb in Allium cepa L. *Cytologia* 84(2): 113–117.

[17] Saquib Q, Siddiqui M A, Ansari S M, Alwathnani H A, Musarrat J, Al-Khedhairy A A. 2021.Cytotoxicity and genotoxicity of methomyl, carbaryl, metalaxyl, and pendimethalin in human umbilical vein endothelial cells. *J. Appl. Toxicol.*, May;41(5):832-846. doi: 10.1002/jat.4139. Epub 2021 Jan 11. PMID: 33427323.

[18] Kamata S, Hashiyama R, Hana-ika H, Ohkubo I, Saito, Honda A, Anan Y, Akahoshi N, Noguchi N, Kanda Y, Ishii I. 2020. Cytotoxicity comparison of 35 developmental neurotoxicants in human induced pluripotent stem cells (iPSC), iPSC-derived neural progenitor cells, and transformed cell lines. *Toxicology In Vitro*, 69, 104999. https://doi.org/10.1016/j.tiv.2020.104999.

[19] Uğurlu P, Satar E İ, Çiçek T. (2019). The histopathological, cytopathological and ultrastructural effects of carbaryl on gills of Oreochromis niloticus (Linnaeus, 1758). *Environmental Toxicology and Pharmacology*, 71, 103217.

[20] Atasayar Ünver Z., Koldemir Gündüz M., Kayhan F. E., Kaymak G., Zengin Ü., Çağatay P., Süsleyici Duman B. (2015) The Effects of

Peroxiredoxin 6 Gene rd41055489 Variation on Oxidative Stress Mechanism in Human Model Organism. Bezmialem Science, 3:37-42.

[21] Aydemir N, Bilaloğlu R. 2004. The investigation of the genotoxic effects of fenarimol and propamocarb in mouse bone marrow in vivo, *Toxicology Letters*, 147, 1, 73-78, ISSN 0378-4274, https://doi.org/10.1016/j.toxlet.2003.10.015.

[22] Yoon, J Y, Oh, S H, Yoo, S M, Lee, S J, Lee, H S, Choi, S J, & Lee, B H (2001). N-nitrosocarbofuran, but not carbofuran, induces apoptosis and cell cycle arrest in CHL cells. *Toxicology*, 169(2), 153-161.

[23] Lee B-W, Oha S-H, Chung J-H, Moonb C-K, Lee B-H. 2004. NNitroso metabolite of carbofuran induces apoptosis in CHL cells by cytochrome c-mediated activation of caspases. *Toxicology* 201:51–8.

[24] Zeljezic D, Vrdoljak AL, Kopjar N, Radic B, Kraus SM. 2008. Cholinesterase-Inhibiting and genotoxic effects of acute carbofuran intoxcitaion in man: A case report. *Basic & Clinical Pharmacology & Toxicology*, 103, 329–335.

[25] Sun X Y, Jin Y T, Wu B, et al. [Study on genotoxicity of aldicarb and methomyl]. Huan Jing ke Xue= Huanjing Kexue. 2010 Dec;31(12):2973-2980. PMID: 21360888.

[26] Doğanyiğit İ, Öztürk Küp F. Effects of Bendiocarb Insecticide on Lipid Peroxidation, Antioxidant Enzymes and DNA Damage in Human Leukocytes. *Erciyes University Journal of Institue Of Science and Tecnology*, 37, 2, 348-355.

Chapter 7

Use of Gene Polymorphisms as Biomarkers in Long-Term Carbamate Use

Meliha Koldemir-Gündüz*
Department of Engineering Basic Sciences, Kütahya Health Sciences University, Kütahya, Turkey

Introduction

Pesticides are used to protect plants from hazardous organisms by controlling, destroying, and preventing pests (insects, weeds, birds, rodents, nematodes and microorganisms) [1]. Pesticides are chemicals that are used to keep pests at bay. Pesticides are categorized as inorganic, synthetic and biological. Organophosphate, pyrethroid, organochlorine, and carbamate are the most frequent pesticide derivatives (CB). These substances are also classified as rodenticide, herbicide, fungicide and insecticide. Carbamates are a form of pesticide that has been around for a long time in agriculture.

According to their chemical structures and biological effects, carbamate pesticides are split into two groups: N-methyl-carbamate insecticides and N-allyl-carbamate herbicides [2-3]. Carbamates are related to the mechanism of action and resistance of organophosphorus compounds and act as inhibitors of acetylcholinesterase. They are toxic to organisms due to their acute toxic effects [4]. Carbamate insecticides have been designated as poisonous and dangerous by the World Health Organization, and their usage has been restricted. Carbamate use is a major source of concern, not only because of its ubiquitous use, but also because it has substantial harmful effects on plants and animals [5]. These chemicals are widely used in the control of insects and

* Corresponding Author's Email: meliha.koldemirgunduz@ksbu.edu.tr

In: The Science of Carbamates
Editor: Güllü Kaymak
ISBN: 978-1-68507-708-2

pests that damage plant crops. Carbamate pesticides can be converted into a variety of chemicals by a variety of mechanisms, including biotransformation, oxidation, biological growth, hydrolysis, photolysis, biological decomposition, and metabolic activities in living organisms [6].

Despite the fact that carbamate insecticides do not stay in aquatic environments for lengthy periods of time and are particularly stable in these environments, their use can result in a significant reduction in non-target creatures. Because of their tendency to damage soil and water, people are becoming more aware of the negative impacts of these pesticides [7].

In recent years, pesticide use has increased significantly all over the world. Carbamates (CBs) are a major source of concern due to their widespread use and negative public health consequences [8-9]. Due to their direct contact to pesticides, agricultural field workers are sensitive to pesticide poisoning [10-11]. The majority of pesticide applicators in agricultural areas are unaware of pesticide's negative effects and frequently do not utilize personal protective equipment. Acute CB exposure might result in dizziness, nausea, vomiting, respiratory difficulties, and even death [12]. The amount of exposure and the individual's health determine the severity of these effects. Nausea, vomiting, runny nose, abdominal pain, dizziness, headache, exhaustion, restlessness, depression, irritability, anxiety, sleep disturbance, somnolence, numbness, paresthesia, limb pain, and limb weakness can all be symptoms of chronic pesticide exposure [13]. Chronic toxicity is determined by a history of employment as a pesticide applicator [14].

CB acts as acetyl cholinesterase (AChE) inhibitors, affecting various organs such as the nervous systems (peripheral and central), liver, pancreas, and muscle. Disruption of enzymatic pathways that modulate biochemical metabolism in the cytosol, mitochondria and peroxisomes may result in AChE inhibition in target organs [15-17]. Pesticides' biological effects are linked to oxidative stress, epigenetic regulation, and genetic polymorphism modulation [18].

Genetic analyzes are molecular biology-based analyzes and are usually performed with blood or tissue belonging to the individual. The most basic use of gene analysis is in the field of medicine. With gene analysis, pre-diagnosis of diseases or predisposition to the disease can be determined. Most diseases occur with the combined effects of genetic and environmental factors. The individual, whose susceptibility to the disease is determined as a result of genetic analysis, can prevent/postpone the disease or increase the chance of treatment by regulating environmental factors. Even if individuals regulate environmental factors and use drugs, sometimes they may not be able to prevent the disease. The most important reason for this situation is the polymorphic differences in the genome.

Organisms of the same species often show different phenotypes from each other. This difference is genetically determined and named as polymorphism. A genetic polymorphism is an allele that has a frequency of more than 1% in a population of two or more alleles at a certain locus. Single nucleotide polymorphisms (SNPs) are referred to as single nucleotide differences in DNA between individuals. The existence of specific polymorphisms in genes encoding DNA repair enzymes or phase I/II detoxification enzymes may cause individual variances in xenobiotic reaction. The appearance of specific allelic variations can either safeguard or make a person more vulnerable to the impacts of xenobiotics. The reason for the different reflections of polymorphisms on the phenotype is due to the existence of multiple alleles. Some SNP alleles are sequence variants that cause a difference in the function or regulation of the gene that directly causes the disease. Genetic variation in genes encoding DNA repair enzymes is linked to the ability to repair DNA. As proteins encoded by different genotypes modify substrate biotransformation, genetic variations may be linked to differing degrees of risk in the general population. Different types of SNPs can alter protein function or gene expression. Although multiple SNPs may directly affect one or several phenotypes, they may not be associated with disease. In this regard, differentiated investigations on bigger samples, even for ethnic groupings, might be beneficial.

In this section, biomarkers that help determine the management of chronic pesticide toxicity by determining different gene polymorphisms in agricultural workers as a result of long-term contact with carbamate groups will be discussed.

Different Gene Polymorphisms and Carbamate Relationship

Polymorphism of CYP2D6 Gene G1934A (rs 3892097)

Pesticides are substances used to control pests. Pesticides are metabolized primarily by several xenobiotic metabolizing enzymes, including cytochrome P450 enzymes. In pesticide exposure, Cytochrome P450 polymorphisms are common biomarkers for assessing susceptibility [19]. The G1934A (rs 3892097) gene polymorphism in the CYP2D6 gene results in a lower catalytic rate of enzyme activity [20]. Pitaksilp et al. In this study, investigating

CYP2D6 G1934A gene polymorphism together with occupational health risks and behaviors in Thai rice growers; and evaluated the link between gene polymorphisms and serum cholinesterase (SChE) levels [21]. Herbicides are the most prevalent pesticides imported into Thailand, followed by insecticides and fungicides [22] Thai farm laborers said they used a range of pesticides and treated them three to four times a month on average [23-24]. Agricultural workers have claimed that abamectin, chloropyrifos, carbofuran, and cypermethrin are the most active insecticidal compounds in commercial treatments they employ for "management of plant diseases" [25-26].

CYP2D6 is a key member of the cytochrome P450 family that works with a wide range of xenobiotic substrates and because of genetic polymorphism, there is a lot of variation in gene expression and enzymatic activity [22]. The isocarbophos desulfuration secondary metabolic pathway is linked to the CYP2D6 gene. Acetylcholinesterase inhibitors have been identified for isocarbofos enantiomers and its oxidative desulfuration metabolite, isocarbofos oxone [27-28]. When compared to control, CYP2D6 inhibitors reduce cholinesterase activity by 30% for chlorpyrifos, 38% for diazinon, and 50% for parathion [29], and SChE is markedly reduced during chronic exposures [30]. CYP2D6 allele variants in the third intron and four exons cause transcriptional loss and single-based mRNA deletion [31]. In an infrequently reported Thai population, the GA genotype was more vulnerable to pesticides than the wild type. The CYP2D6 (G1934A) polymorphism may be a useful biomarker in determining the management of chronic pesticide toxicity in combination with SChE activity testing for job assignments, particularly in agricultural workers, particularly in the risk group.

GST M1 and GST T1 Polymorphisms

The biomarkers AChE and plasma Butyrylcholinesterase (BChE) are commonly employed to measure CB pesticide exposures [9, 32]. AChE regulates nerve impulse propagation by converting acetylcholine to acetic acid and choline [33]. BChE is renowned for its ability to detoxify xenobiotics, as well as its impact on acetylcholine hydrolysis and lipid metabolism [34]. Glutathione S-Transferases (GSTs) are a family of enzymes that use glutathione to conjugate endogenous and exogenous molecules like electrophilic compounds, carcinogens, medicinal medicines, environmental pollutants, and oxidative stress products for detoxification [35]. There are eight cytosolic GST enzyme classes in humans [36]. Because of their role in drug metabolism, the Theta, Mu, Alpha, and Pi classes have gotten greater attention than other classes [36]. In the human population, polymorphisms in GST T1 and GST M1 are common [37]. In the human population, the GST M1 and GST T1 polymorphisms are widespread. Four genotypic individuals are feasible based on the presence of GST M1 and GST T1: those who are both positive for GST M1 and GST T1; GST M1 positive and GST T1 null; GST M1 null and GST T1 positive; and GSTM1 and GST T1 are both null. Gene deletion is the primary cause of GST M1 and GST T1 polymorphism. GST M1 bladder [38], lung [39], skin tumors [40], and colon cancer [41] null genotype. While the GST T1 genotype has been connected to an elevated risk of cancer in the oral cavity, head, and neck area, no other cancers have been linked to it [42].

CBs have been shown to reduce butyrylcholinesterase (BChE) and acetylcholinesterase (AChE) activity while increasing Glutathione S-Transferase (GST) activity in people who have been exposed to them. In a study conducted by Dutta and colleagues in Lower Himalayan West Bengal, BChE and AChE activities were much lower than in the working group compared to the control group in a study that evaluated BChE, AChE, and GST activities in tea garden workers exposed to pesticides for a long time and attempted to correlate enzyme activities with polymorphisms in the GST M1 and GST T1 genes [43]. The worker group's GST activity was found to be substantially higher than the control group's. A total of 122 people were tested, with 34.43 percent having both GST M1 and GST T1, 23.77 percent having a single null genotype for GST T1, 27.05 percent having a single null genotype for GST M1, and 14 percent having both GST M1 and GST T1. 75 of the participants had double null genotypes. When the four genotypes were compared, there was no significant difference in enzyme activity between the worker and control groups [43]. While the presence or absence of GST M1 and GST T1 alleles was linked to changes in AChE, BChE, and GST activities in tea garden workers who were chronically exposed to CBs in the study area, the presence or absence of GST M1 and GST T1 alleles was not linked to changes in AChE, BChE, and GST activities [43]. According to information from different studies, GST M1 and GST T1 null genotypes are associated with pesticide-related symptoms in individuals with long-term pesticide exposure [32]. Pesticide exposure has been associated to an increased risk of cutaneous melanoma when combined with the GST M1 null genotype [44], and pesticide exposure has been associated to an increased risk of Parkinson's disease when combined with the GST M1 and GST T1 null genotypes [45].

GSTP1 and XRCC1 Polymorphisms

Pesticide exposure has been related to an increased risk of cancer and genotoxicity. Pesticides' glutathione transferase-dependent metabolism, as well as polymorphisms in DNA repair genes, may influence these hazards [46]. The "International Agency for Cancer Research" (IARC) has established that some pesticides pose risks to human health [47-48]. Carbamates are one of the substances linked to negative health effects (49), and biomarkers can be used to predict occupational exposure [50-51]. Many DNA repair pathways minimize xenobiotic genotoxicity [48]. The gene XRCC1 is involved in DNA repair that has been shown to defend against pesticide-induced genotoxicity [49]. In pesticide-exposed populations, changes in exon 6 (Arg194Trp) (CT) and exon 10 (Arg399Gln) (GA) of the XRCC1 gene have been related to a greater risk of DNA damage [52].

Previous research has found that workers exposed to various pesticide mixes suffer from higher DNA damage [53]. Polymorphisms in metabolism and DNA repair genes can alter the degree of DNA damage in pesticide-exposed workers [48]. GST enzymes play a significant role in pesticide detoxification [54]. In a study looking at the prevalence of GSTP1 and XRCC1 polymorphisms and their possible correlation with DNA damage after long-term pesticide exposure, Saad-Hussein and colleagues discovered that workers exposed to pesticides with the GSTP1 Ile-Ile genotype had more DNA damage than workers with the other two genotypes (Ile-Val and Val-Val) [46]. These findings matched those of other studies [48, 55], which discovered a rise in DNA damage in the GSTP1 gene Pesticide combination workers with the Ile-Ile genotype. The GSTP1 Ile-Ile genotype was found to be strongly related with higher DNA damage in another study [54], particularly in those with heavy pesticide exposure. However, some research [56-57] has revealed no link between DNA damage and GSTP1 genotypes. Cells are protected from genotoxicity through DNA repair. The XRCC1 protein preserves genetic stability by repairing DNA strand breaks [52]. Saad-Hussein et al. discovered that the XRCC1 399Arg-Arg genotype was connected to increased DNA damage in pesticide-exposed workers, despite finding no link between DNA damage and the XRCC1194 gene polymorphism [46].

NQO1 Polymorphism

Pesticides are widely used in agriculture, posing a harm to those who are exposed to them. Pesticide poisoning in the workplace affects a vast number of individuals, particularly in developing nations where agriculture employs a large percentage of the workforce and pesticide application occurs under unfavorable conditions, such as the use of restricted compounds and improper spray equipment [58-59]. A rising amount of evidence suggests a link between occupational pesticide exposure and the development of a wide range of diseases, from diabetes to cancer and neurological diseases [60]. Because the metabolism of certain pesticides might result in the production of "Reactive Oxygen Species" (ROS), polymorphisms in genes encoding antioxidant enzymes appear to have an important role. "Quinone oxidoreductase 1" (NQO1) is a cytosolic enzyme that catalyzes the two-electron reduction of quinoids with NAD(P)H as a cofactor, detoxifying electrophilic chemicals and preventing the generation of ROS [61]. Because it affects the enzyme's catalytic activity, the nucleotide substitution cytosine (C) to thymine (T) at position 609 of the NQO1 gene, which codes for the conversion of proline to serine at codon 187 (rs1800566), is noteworthy. Homozygous mutant variation (TT) and heterozygous variant genotype (CT) were shown to have lower enzyme activity, which could have an additive effect on oxidative damage [62].

Ahmadi et al. [63] reported that the quinone oxidoreductase 1 (NQO1) C609T polymorphism was not substantially linked with the biomarkers investigated in greenhouse workers occupationally exposed to pesticides. While correcting for characteristics such as smoking status, body mass index, age, and NQO1 polymorphism, Zepeda-Arce et al. [64] showed no association between multivariate linear regression analysis and the NQO1 polymorphism rs1800566 of pesticide exposure as a predictor.

Conclusion

Biological effects are known to be regulated by the formation of genetic polymorphisms, despite the fact that oxidative stress and epigenetic alterations are the principal pathogenic mechanisms generated by pesticide exposure. Under the effect of genetic polymorphisms, certain people are more sensitive to pesticide-induced oxidative stress [65]. Individuals exposed to specific pollutants may be more susceptible to the development of chronic/acute disorders as a result of genetic inheritance [66-69]. Genetic differences can alter individual exposure to xenobiotics like insecticides. More research into the diversity of DNA repair genes and their combinations is required to determine their true role as pesticide toxicity drivers. DNA repair and xenobiotic metabolism genes may hold the key to better understanding pesticide-induced sickness. The development of screening approaches for large-scale detection of pesticide or other xenobiotic susceptibility could be aided by a complete understanding of these complex pathways. In the near future, improving preventive measures to safeguard in terms of both occupational and environmental hazards, they are the most vulnerable contexts might be a top focus [65].

References

[1] Umar Mustapha, M., Halimoon, N., Lutfi, W., Johari, W., Yunus, M., Shukor, A., Halimoon, N., Johar, W., Yunus, M. (2019). An Overview on Biodegradation of Carbamate Pesticides by Soil. *Bacteria. Pertanika J. Sci. & Technol.* 27, 547-563.

[2] Parks, P., Lipman, J., Eidelman, J. (1987) Carbamate Toxicity - a Case-Report. South African. *Medical Journal*, 72, 222-222.

[3] Ozturk, B., Ghequire, M., Nguyen, T. P. O., De Mot, R., Wattiez, R., Springael, D. (2016). Expanded insecticide catabolic activity gained by a single nucleotide substitution in a bacterial carbamate hydrolase gene. *Environmental Microbiology*, 18, 4878-4887.

[4] Alvarez, A., Saez, J. M., Costa, J. S. D., Colin, V. L., Fuentes, M. S., Cuozzo, S. A., Benimeli, C. S., Polti, M. A., Amoroso, M. J. (2017) Actinobacteria: Current research and perspectives for bioremediation of pesticides and heavy metals. *Chemosphere,* 166, 41-62.

[5] Kaymak, G. (2020). Experimental evidence of the sublethal effects of carbamate pesticide on zebrafish (*Danio rerio*). *Fresenius Environmental Bulletin,* 29(9), 7267-7274.

[6] Cai, Z. Q., Wang, J., Ma, J. T., Zhu, X. L., Cai, J. Y., Yang, G. H. (2015). Anaerobic Degradation Pathway of the Novel Chiral Insecticide Paichongding and Its Impact on Bacterial Communities in Soils. *Journal of Agricultural and Food Chemistry,* 63, 7151-7160.

[7] Kocak B. (2020) An overview of behaviours of carbofuran in soil. 4. *Cukurova International Scientific Research Congress*, 21-23 February Adana 106-114.

[8] Quandt, S. A., Pope, C. N., Chen, H., Summers, P., Arcury, T. A. (2015). Longitudinal assessment of blood cholinesterase activities over 2 consecutive years among Latino nonfarmworkers and pesticide-exposed farmworkers in North Carolina. *J. Occup. Environ. Med.* 57, 851-857.

[9] Cotton, J., Edwards, J., Rahman, M. A., Brumby, S. (2018). Cholinesterase research outreach project (CROP): point of care cholinesterase measurement in an Australian agricultural community. *Environ. Health* 17, 31.

[10] Farahat, F. M., Ellison, C. A., Bonner, M. R., McGarrigle, B. P., Crane, A. L., Fenske, R. A., Lasarev, M. R., Rohlman, D. S., Anger, W.K., Lein, P. J., Olson, J. R. (2011). Biomarkers of chlorpyrifos exposure and effect in Egyptian cotton field workers. *Environ. Health Perspect.* 119, 801-806.

[11] Garabrant, D. H., Aylward, L. L., Berent, S., Chen, Q., Timchalk, C., Burns, C. J., Hays, S. M., Albers, J. W. (2009). Cholinesterase inhibition in chlorpyrifos workers: characterization of biomarkers of exposure and response in relation to urinary TCPy. *J. Expo. Sci. Environ. Epidemiol.* 19, 634-642.

[12] Strelitz, J., Engel, L. S., Keifer, M. C. (2014). Blood acetylcholinesterase and butyrylcholinesterase as biomarkers of cholinesterase depression among pesticide handlers. *Occup. Environ. Med.* 71, 842-847.

[13] Jinatana, S., Sming, K., Krongtong, Y., Thanyachai, S. (2009). Cholinesterase activity, pesticide exposure and health impact in a population exposed to organophosphates. *Int. Arch. Occup. Environ. Health,* 82, 833-842.

[14] Lee, B. W., London, L., Paulauskis, J., Myers, J., Christiani, D. C. (2003). Association between human paraoxonase gene polymorphism and chronic symptoms in pesticide-exposed workers. *J. Occup. Environ. Med.* 45, 118-122

[15] Pohanka, M. (2020). Inhibitors of cholinesterases in the pharmacology, the current trends. *Mini Rev Med Chem,* 20(15):1532-1542.

[16] King, A. M., Aaron, C. K. (2015). Organophosphate and carbamate poisoning. *Emerg Med Clin North Am.,* 33(1): 133-151.

[17] Karami-Mohajeri, S., Abdollahi, M. (2011). Toxic influence of organophosphate, carbamate, and organochlorine pesticides on cellular metabolism of lipids, proteins, and carbohydrates: a systematic review. *Hum Exp Toxicol,* 30(9): 1119-1140.

[18] Teodoro, M., Briguglio, G., Fenga, C., Costa, C. (2019). Genetic polymorphisms as determinants of pesticide toxicity: Recent advances. *Toxicol Rep.,* 6: 564-570.

[19] Kapka-Skrzypczak, L., Cyranka, M., Skrzypczak, M., Kruszewski, M. (2011). Biomonitoring and biomarkers of organophosphate pesticides exposure – state of the art. *Ann Agric Environ Med.* 18: 294-303.

[20] Singh, S., Kumar, V., Vashisht, K., Singh, P., Banerjee, B. D., Rautela, R. S., Grover, S. S., Rawat, D. S., Pasha, S. T., Jain, S. K., Rai, A. (2011). Role of genetic polymorphisms of CYP1A1, CYP3A5, CYP2C9, CYP2D6, and PON1 in the modulation of DNA damage in workers occupationally exposed to organophosphate pesticides. *Toxicol Appl Pharmacol.*, 257(1): 84-92.

[21] Pitaksilp, T., Ruchiwit, K., Suwannahong, K., Pongstaporn, W., Sudjaroen, Y. (2020). CYP2D6 (G1934A) Gene Polymorphism in Rice Farmers with Long-term Pesticide Exposure, Suphan Buri, Thailand. *Indian Journal of Forensic Medicine & Toxicology, October-December,* 14, *3* 2678-2684.

[22] Ministry of Agricultural and Cooperatives Office of Agricultural Economics. Bangkok: [homepage on the Internet]. *Summary of imported pesticides (in Thai).* Available from: www.oae.go.th/ewt_news.php?nid=146&filename=index.

[23] Panuwet, P., Prapamontol, T., Chantara, S., Thavornyuthikarn, P., Montesano, M. A. Whitehead, R. D. Jr, Barr, D. B. (2008). Concentrations of urinary pesticide metabolites in small-scale farmers in Chiang Mai Province, *Thailand. Sci Total Environ.,* 407(1): 655-668.

[24] Kachaiyaphum, P., Howteerakul, N., Sujirarat, D., Siri, S., Suwannapong, N. (2010). Serum cholinesterase levels of Thai chili-farm workers exposed to chemical pesticides: prevalence estimates and associated factors. *J Occup Health.*, 52(1): 89-98.

[25] Prasertsung, N. (2012). Situation of pesticides used in rice fields in Suphanburi Province; *Proceedings of the conference on chemical pesticides.* Nov 15-16. Available from: www.thaipan.org/sites/default/files/conference2555/conference2555_1_10.pdf.

[26] Sapbamrer, R. (2018). Pesticide use, poisoning, and knowledge and unsafe occupational practices in Thailand. *New Solutions,* 28(2): 283-302.

[27] Kaur, G., Jain, A. K., Singh, S. (2017). CYP/PON genetic variations as determinant of organophosphate pesticides toxicity. *J Genet.*, 96(1): 187-201.

[28] Zhuang, X. M., Wei, X., Tan, Y., Xiao, W. B., Yang, H. Y., Xie, J. W., Lu, C., Li, H. (2014). Contribution of carboxylesterase and cytochrome P450 to the bioactivation and detoxification of isocarbophos and its enantiomers in human liver microsomes. *Toxicol Sci.,* 140(1): 40-48.

[29] Sams, C., Mason, H. J., Rawbone, R. (2000). Evidence for the activation of organophosphate pesticides by cytochromes P450 3A4 and 2D6 in human liver microsomes. *Toxicol Lett.,* 116(3): 217-221.

[30] Tawfik Khattab, A. M., Zayed, A. A., Ahmed, A. I., Abdel Aal, A. G., Mekdad, A. A. (2016). The role of PON1 and CYP2D6 genes in susceptibility to organophosphorus chronic intoxication in Egyptian patients. *Neurotoxicology.* 53: 102-107.

[31] Hanioka, N., Kimura, S., Meyer, U. A., Gonzalez, F. J. (1990). The human CYP2D locus associated with a common genetic defect in drug oxidation: a G1934 A base change in intron 3 of a mutant CYP2D6 allele results in an aberrant 3' splice recognition site. *Am J Hum Genet.* 47(6): 994-1001.

[32] Hernandez, A. F., López, O., Rodrigo, L., Gil, F., Pena, G., Serrano, J. L., Parrón, T., Alvarez, J. C., Lorente, J. A., Pla, A. (2005). Changes in erythrocyte enzymes in humans long-term exposed to pesticides Influence of several markers of individual susceptibility. *Toxicol. Lett.* 159, 13-21.

[33] Mwila, K., Burton, M. H., Van Dyk J. S., Pletschke, B. I. (2013). The effect of mixtures of organophosphate and carbamate pesticides on acetylcholinesterase and application of chemometrics to identify pesticides in mixtures. *Environ. Monit. Assess.* 185, 2315-2327.

[34] Lockridge, O. (2015). Review of human butyrylcholinesterase structure, function, genetic variants, history of use in the clinic, and potential therapeutic uses. *Pharmacol. Ther.,* 148, 34-46.

[35] Nebert, D. W., Vasiliou, V. (2004). Analysis of the glutathione S-transferase (GST) gene family. *Hum. Genomics.* 1, 460-464.

[36] Josephy, P. D. (2010). Genetic Variations in Human Glutathione Transferase Enzymes: Significance for Pharmacology and Toxicology. *Hum. Genomics Proteomics* 876940.

[37] Girisha, K. M., Gilmour, A., Mastana, S., Singh, V. P., Sinha, N., Tewari, S., Ramesh, V., Sankar, V. H., Agrawal S. (2004). T1 and M1 polymorphism in Glutathione S-transferase gene and coronary artery disease in North Indian population. *Indian J. Med. Sci.* 58, 520-526.

[38] Brockmoller, J., Kerb, R., Drakoulis, N., Staffeldt, B., Roots, I. (1994). Glutathione S-Transferase Ml and Its Variants A and B as Host Factors of Bladder Cancer Susceptibility: A Case-Control Study. *Cancer Res.* 54, 4103-4111.

[39] Sreeja, L., Syamala, V., Hariharan, S., Madhavan, J., Devan, S. C., Ankathil, R. (2005). Possible risk modification by CYP1A1, GSTM1 and GSTT1 gene polymorphisms in lung cancer susceptibility in a South Indian population. *J. Hum. Genet.,* 50, 618-627.

[40] Heagerty, A. H., Fitzgerald, D., Smith, A., Bowers, B., Jones, P., Fryer, A. A., Zhao, L., Alldersea, J., Strange, R. C. (1994). Glutathione Stransferase GSTM1 phenotypes and protection against cutaneous tumors. *Lancet* 343, 266-268.

[41] Lohmueller, K. E., Pearce, C. L., Pike, M., Lander, E. S., Hirschhorn, J. N. (2003). Meta-analysis of genetic association studies supports a contribution of common variants to susceptibility to common disease. *Nat. Genet.* 33, 177-182.

[42] Strange, R. C., Fryer, A. A. (1999). The glutathione S-transferases: influence of polymorphism on cancer susceptibility. *IARC Sci. Publ.* 148, 231-249.

[43] Dutta, T., Nayak, C., Bhattacharjee, S. (2019). Acetylcholinesterase, Butyrylcholinesterase and Glutathione S-Transferase Enzyme Activities and Their Correlation with Genotypic Variations Based on GST M1 and GST T1 Loci in Long Term-Pesticide-Exposed Tea Garden Workers of Sub-Himalayan West Bengal. *Toxicol. Environ. Health. Sci.,* 11(1), 63-42.

[44] Fortes, C., Mastroeni, S., Bottà, G., Boffetta, P., Antonelli, G., Venanzetti, F. (2016). Glutathione S-transferase M1 null genotype, house hold pesticides exposure and cutaneous melanoma. *Melanoma Res.,* 26, 625-630.

[45] Pinhel, M. A. de S., Sado, C. L., Longo, G. Dos S., Gregório, M. L., Amorim, G. S., Florim, G. M. da S., Mazeti, C. M., Martins, D. P., Oliveira, F. de N., Nakazone, M. A., Tognola, W. A., Souza, D. R. S. (2013). Nullity of GSTT1/GSTM1 related to pesticides is associated with Parkinson's disease. *Arq. Neuropsiquiatr* 71, 527-532.

[46] Saad-Husseina, A., Noshyb, M., Tahaa, M., El-Shorbagyb, H., Shahya, E., Abdel-Shafya, E. A. (2017). GSTP1 and XRCC1 polymorphisms and DNA damage in agricultural workers exposed to pesticides. *Mutat Res Gen Tox En.,* 819, 20–25.

[47] Das, P. P., Shaik, A. P., Jamil, K. (2007). Genotoxicity induced by pesticide mixtures: in-vitro studies on human peripheral blood lymphocyte. *Toxicol. Ind. Health,* 23, 449–458.

[48] Wong, R. H., Chang, S. Y., Ho, S. W., Huang, P. L., Liu, Y. J., Chen, Y. C., Yeh, Y. H., Lee, H. S. (2008). Polymorphisms in metabolic GSTP1 and DNA-repair XRCC1 genes with an increased risk of DNA damage in pesticide-exposed fruit growers. *Mutat. Res.,* 654, 168–175.

[49] Kapka-Skrzypczak, L., Cyranka, M., Skrzypczak, M., Kruszewski, M. (2011). Biomonitoring and biomarkers of organophosphate pesticides exposure– state of the art. Ann. Agric. Environ. Med., 18, 294–303.

[50] Barr, D. B., Bravo, R., Weerasekera, G., Caltabiano, L. M., Whitehead Jr, R. D., Olsson, A. O., Caudill, S. P., Schober, S. E., Pirkle, J. L., Sampson, E. J., Jackson, R. J., Needham, L. L. (2004). Concentrations of dialkyl phosphate metabolites of organophosphorus pesticides in the U. S. Population. *Environ. Health Perspect.* 112, 186–200.

[51] Ngo, M. A., O'Malley, M., Maibach H. I. (2010). Percutaneous absorption and exposure assessment of pesticides. *J. Appl. Toxicol,.* 30, 91–114.

[52] Fan, J., Otterlei, M., Wong, H. K., Tomkinson, A. E., Wilson, D. M. (2004). XRCC1 co-localizes and physically interacts with PCNA. *Nucleic Acids Res.,* 32, 2193–2201.

[53] Atherton, K. M., Williams, F. M., Egea González, F. J., Glass, R., Rushton, S., Blain, P. G., Mutch, E. (2009). DNA damage in horticultural farmers: a pilot study showing an association with organophosphate pesticide exposure. *Biomarkers, 14, 443–451.*

[54] Liu, Y. J., Huang, P. L., Chang, Y. F., Chen, Y. H., Chiou, Y. H., Xu, Z. L., Wong, R. H. (2006). GSTP1 genetic polymorphism is associated with a higher risk of DNA damage in pesticideexposed fruit growers. *Cancer Epidemiol. Biomarkers Prev*., 15, 659–666.

[55] Singh, S., Kumar, V., Singh, P., Thakur, S., Banerjee, B. D., Rautela, R. S., Grover, S. S., Rawat, D. S., Pasha, S. T., Jain, S. K., Rai, A. (2011). Genetic polymorphisms of GSTM1, GSTT1 and GSTP1 and susceptibility to DNA damage in workers occupationally exposed to organophosphate pesticides. *Mutat. Res.,* 725, 36–42.

[56] Da Silva, J., Moraes, C. R., Heuser, V. D., Andrade, V. M., Da Silva, F. R., Kvitko, K., Emmel, V., Rohr, P., Bordin, D. L., Andreazza, A. C., Salvador, M., Henriques, J. A., Erdtmann, B. (2008). Evaluation of genetic damage in a Brazilian population occupationally exposed to pesticides and its correlation with polymorphisms in metabolizing genes. *Mutagenesis,* 23, 415–422.

[57] Da Silva, F. R., Da Silva, J., Mda Allgayer, C., Simon, C. F., Dias, J. F., dos Santos, C. E., Salvador, M., Branco, C., Schneider, N. B., Kahl, V., Rohr, P. (2012). Genotoxic biomonitoring of tobacco farmers: biomarkers of exposure, of early biological effects and of susceptibility. *J. Hazard. Mater.,* 225-226, 81–90.

[58] Colosio, C., Vellere, F. (2010). Epidemiological studies of anticholinesterase pesticide poisoning: Global impact. In *Anticholinesterase Pesticides: Metabolism, Neurotoxicity, and Epidemiology*; Satoh, T., Gupta, R. C., Eds.; John Wiley & Sons, Inc.: Hoboken, NJ, USA, pp. 343–356. ISBN 9780470410301.

[59] Damalas, C. A., Eleftherohorinos, I. G. (2011). Pesticide exposure, safety issues, and risk assessment indicators. *Int. J. Environ. Res. Public Health,* 8, 1402–1419.

[60] Gangemi, S., Miozzi, E., Teodoro, M., Briguglio, G., De Luca, A., Alibrando, C., Polito, I., Libra, M. (2016). Occupational exposure to pesticides as a possible risk factor for the development of chronic diseases in humans. *Mol. Med. Rep.,* 14, 4475–4488.

[61] Ross, D., Kepa, J. K., Winski, S. L., Beall, H. D., Anwar, A., Siegel, D. (2000). Nad(p)h:Quinone oxidoreductase 1 (NQO1): Chemoprotection,

bioactivation, gene regulation and genetic polymorphisms. *Chem. Biol. Interact.* 129, 77–97.

[62] Boroumand, M., Pourgholi, L., Goodarzynejad, H., Ziaee, S., Hajhosseini-Talasaz, A., Sotoudeh-Anvari, M., Mandegary, A. (2017). NQO1 C609T polymorphism is associated with coronary artery disease in a gender-dependent manner. *Cardiovasc. Toxicol.* 17, 35–41.

[63] Ahmadi, N., Mandegary, A., Jamshidzadeh, A., Mohammadi-Sardoo, M., Mohammadi-Sardo, M., Salari, E., Pourgholi, L. (2018). Hematological Abnormality, Oxidative Stress, and Genotoxicity Induction in the Greenhouse Pesticide Sprayers; Investigating the Role of NQO1 Gene Polymorphism. *Toxics*, 6, 13 1-15.; doi:10.3390/toxics6010013.

[64] Zepeda-Arce, R., Rojas-García, A. E., Benitez-Trinidad, A., Herrera-Moreno, J. F., Medina-Díaz, I. M., Barrón-Vivanco, B. S., Villegas, G. P., Hernández-Ochoa, I., Sólis Heredia, M. d. J., Bernal-Hernández, Y. Y. (2017). Oxidative stress and genetic damage among workers exposed primarily to organophosphate and pyrethroid pesticides. *Environ. Toxicol.*, 32, 1754–1764.

[65] Teodoroa, M., Briguglioa, G., Fengaa, C., Costab, C. (2019). Genetic polymorphisms as determinants of pesticide toxicity: Recent Advances. *Toxicology Reports,* 6, 564–570.

[66] Kaur, G., Jain, A. K., Singh, S. (2017). CYP/PON genetic variations as determinant of organophosphate pesticides toxicity. *J. Genet.* 96, 187–201.

[67] Costa, C., Gangemi, S., Giambo, F., Rapisarda, V., Caccamo, D., Fenga, C. (2015). Oxidative stress biomarkers and paraoxonase 1 polymorphism frequency in farmers occupationally exposed to pesticides. *Mol. Med. Rep.*, 12, 6353–6357.

[68] Costa, C., Miozzi, E., Teodoro, M., Fenga, C. (2019). Influence of genetic polymorphism on pesticide-induced oxidative stress. *Curr. Opin. Toxicol.*, 13, 1–7.

[69] Saad-Hussein, A., Noshy, M., Taha, M., El-Shorbagy, H., Shahy, E., Abdel-Shafy, E. A. (2017). GSTP1 and XRCC1 polymorphisms and DNA damage in agricultural workers exposed to pesticides. *Mutat. Res. Toxicol. Environ. Mutagen* 819, 20–25.

Chapter 8

Carbamate Intoxication with Clinical Data

Meliha Koldemir-Gündüz
Department of Engineering Basic Sciences, Kütahya Health Sciences University, Kütahya, Turkey

Introduction

Carbamates are carbamic acid compounds with low water solubility and low vapor pressure. Therefore, they evaporate at room temperature, disperse in air and decompose in waters. Carbamates that enter the soil are metabolically degraded. Due to their cholinergic properties, carbamic acid esters are used in pharmacology [1]. Carbamate insecticides are hazardous to animals and humans as well as to the target organism. The most typical adverse effects are in humans, poisoning by carbamate pesticides is mainly caused by inhibition of acetylcholinesterase (AChE). Compared to other pesticides, this inhibition is reversible, and the adverse effect is minimal. Metabolic biotransformation and detoxification pathways may help reduce toxic effects. Lacrimation, increased salivation, miosis, convulsions, constriction of the pupils, high-grade of concussion, and death are the most common signs of poisoning by carbamic acid esters. This takes roughly an hour in acute poisoning. Carbamates react with nitrites to create mutagenic and carcinogenic nitrogen derivatives that cause toxicity, despite their detoxifying and modest mutagenic effects [2].

Carbamate poisoning is widespread in developing nations, where easy access to pesticides and sales limits are insufficient. Carbamate poisonings have a mortality rate of 3–25 percent, depending on the substance consumed, the amount, the patient's previous health status, time spent in the environment, time required for intervention, respiratory support, intubation, and weaning

In: The Science of Carbamates
Editor: Güllü Kaymak
ISBN: 978-1-68507-708-2

from the ventilator [3]. Measurement of serum acetylcholinesterase level is a useful diagnostic criterion that can be used to predict mortality, severity of the disease, and poor prognosis [4, 5]. The intensity of clinical signs and symptoms often corresponds to the level of AChE activity suppression.

This section discusses clinical information on poisoning that may occur as a result of exposure to carbamate groups.

Epidemiology

AChE-inhibiting pesticide poisoning are believed to be responsible for more deaths than any other medicine or chemical class [6]. This is a significant problem in underdeveloped countries because highly toxic pesticides are widely available and widely employed in the pharmaceutical business [7]. In affluent countries, the disease burden of carbamate poisoning is substantially lower. In contrast to the 25,288 people who committed suicide with pesticides in India in recent years [8], the American Association of Poison Control Centers recieved 4150 calls for carbamate poisoning, resulting in three deaths [9]. Accidental poisonings in agriculture usually have a milder effect [10, 11].

Targets of Carbamates

Carbamates keep insect vectors away from commercial and food crops. In addition, they control insect infestations in both commercial and residential areas. Eliminating insect infestations in humans and animals is one of the medical indications. They are also known as nerve agents and are used as chemical weapons and weapons of mass destruction. Nerve agents always pose a danger. Carbamates are the most commonly used insecticides in terms of quantity. However, there have been a lot of intriguing medical applications discovered. Tacrine (donepezil) for Alzheimer's disease, Pyridostigmine (neostigmine) for reversal of neuromuscular blockade, and Physostigmine (ecthiopath) for glaucoma are the most commonly used of these medications [12].

Toxicological Significance of Carbamates

The widespread use of carbamates in medicine creates the potential for overdose and toxicity that must be adequately treated [13-16]. This is much more essential when we consider the dangers of carbamate poisoning. Given the clinical picture of carbamate poisoning, quantitative aspects of the use of anticholinergics and oximes are still being studied [17]. The use of oximes in the treatment of carbamate poisoning is a contentious topic. In several cases of carbaryl intoxication, Farago et al., discovered that pralidoxime chloride (2-PAM) elevated toxicity fatally [18]. Benfuracarb [19], aldicarb [20], and methomyl [20] toxicity were unaffected by 2-PAMs, and cholinesterase reactivation was unaffected by 2-PAM/obidoxime (LüH-6) as well. Lower dosages of 2-PAM or HI-6 oximes have been demonstrated to diminish carbaryl's LD50 in rodents in several investigations [21, 22]. In carbamate toxicity, Hoffman et al., discovered that 2-PAM administered without atropine antagonizes peripheral nicotinic activity [15].

Pharmacodynamic Mechanisms of Carbamates

Carbamates are rapidly absorbed through food, inhalation, the skin, and the eyes. They are most commonly ingested by children and adolescents through oral exposure, accidents or overdose, and as a result of a suicide attempt [23]. Carbamates inhibit AChE and cholinesterase in a pseudoreversible manner [24]. Due to an excess of acetylcholine accumulating in cholinergic synapses, carbamates cause pharmacological and toxic consequences [25]. Ache is mainly found in nerve tissues and erythrocytes. Plasma cholinesterase (pseudocholinesterase, butyrylcholinesterase) is found in serum, liver, heart, pancreas and brain. Acetylcholine is a neurotransmitter found in the central nervous system, as well as the autonomic and somatic nervous systems. After neurochemical transmission, cholinesterases hydrolyze acetylcholine to its inert components, choline and acetic acid. When the enzyme cholinesterase is inhibited, acetylcholine accumulates at nerve synapses and neuromuscular junctions, resulting in overstimulation of acetylcholine receptors. This initial overstimulation is followed by paralysis of cholinergic synaptic transmission in the central nervous system (CNS), autonomic ganglia, parasympathetic and some sympathetic nerve endings (sweat glands), and somatic nerves. As a result, clinical signs of cholinergic crisis occur [26]. The enzyme cholinesterase is transiently and reversibly inhibited by carbamates. Within minutes or hours, enzyme activity returns to normal. Poisoning symptoms last only a few hours. CNS toxicity is uncommon, and seizures are unlikely, due to carbamate's limited CNS penetration. CNS findings are substantially more common in the pediatric age group. Within 4-8 hours, cholinesterase levels recover to normal [26]. However, poisoning with medical or agricultural carbamates that is not treated rigorously and promptly can result in mortality [27].

Clinical Signs and Findings of Carbamate Poisoning

Carbamate poisoning causes signs and symptoms in the muscarinic and nicotinic cholinergic systems, central nervous system, respiratory system, and cardiovascular system [28]. Both erythrocyte and plasma AChE enzyme activity reduce upon carbamate. The intensity of clinical indications and symptoms often corresponds to the level of AChE activity suppression. If AChE activity is 20-50 percent of normal in acute poisoning, mild, 10-20 percent moderate, and less than 10 percent severe symptoms of poisoning occur. The cause of death in carbamate poisoning is usually respiratory failure due paralysis of the respiratory muscles [28].

The agents, the amount absorbed, and the route of exposure all play a role in the clinical picture. If a sufficient amount of poison is absorbed, symptoms may occur within minutes. Within 8–24 hours, patients usually begin to developsymptoms. With inhalation causes the first signs to appear quickly, while transdermal absorption takes the longest. However, opening of the skin on contact may accelerate transdermal absorption [29].

Carbamate poisoning affects the CNS, causing muscarinic, nicotinic, and somatic-motor changes. The most common CNS symptoms are anxiety, insomnia, tremors, dizziness, headache, mental confusion, and hallucinations [29].

Stimulation of muscarinic receptors in the presence of acetylcholine is usually predominant and causes salivation, lacrimation, sweating, urinary incontinence, diarrhea, gastrointestinal distress, vomiting, and bradycardia. Bronchospasm and bronchorea resulting from excess acetylcholine may cause hypoxia and tachycardia. Miotic pupils, blotchy, and pupillary constriction occur with a cholinergic effect [29, 30].

Pallor, mydriasis, tachycardia, and hypertension are all symptoms of cholinergic overstimulation. Cramping and muscle weakness are caused by nicotinic activation at the neuromuscular junction. In carbamate poisonings, increased bronchial secretions, bronchospasm, and dyspnea can occur, especially in moderate and severe poisonings. Respiratory failure, non-cardiogenic pulmonary edema, and acute lung damage can all occur in severe poisonings [31]. Acute respiratory failure and death may occur if the respiratory muscles are paralyzed [32].

In addition, prolongation or shortening of the prothrombin time, an increase or decrease in Factor VII levels, and leukocytosis may rarely appear be seen in carbamate poisonings [33].

Treatment of Carbamate Poisoning

In case of carbamate poisoning, life-saving emergency treatment and especially antidote therapy should be initiated as soon as possible [34]. The main symptomatic antidote for carbamate poisoning is atropine. Atropine can pass the blood-brain barrier and counteract both central and peripheral muscarinic symptoms in carbamate poisoning due to its lipophilic nature [35].

In carbamate poisoning, first of all, it is primarily significant to protect airway patency, ensure airway patency and maintenance provide and support circulation. Keeping the airway open and providing adequate oxygen are the most important and urgent treatment steps. Other emergencies requiring intervention include seizures, electrolyte imbalances, hypoglycemia, and cardiac arrhythmias [26].

If carbamate poisoning is transmitted through the skin, the patient's clothes should be removed, and the skin should be washed with soapy and warm water. Since even a small amount of carbamate can be easily transmitted and lead to severe poisoning, it is important that medical treating the patient to protect themselves with gloves and masks. In the first hour after oral carbamate ingestion, activated charcoal is administered and gastric lavage is performed if the patient is conscious and not vomiting [34]. Activated charcoal is contraindicated if the airway is unprotected and the digestive system is not functioning properly, as complications such as vomiting and aspiration, may occur and there is a risk of acute respiratory failure that may follow aspiration [36]. Gastric lavage is recommended if toxic ingestion occurs within 1 hour [26].

The muscarinic receptor antagonist atropine is the main antidote with proven efficacy in carbamate poisoning. It is useful in antagonizing the cholinergic manifestations caused by the accumulation of acetylcholine and overstimulation of cholinergic receptors. Atropine should be administered as soon as possible and the dose should be repeated. Intravenous administration of atropine should be continued until atropinization goals are achieved [34].

The amount of atropine required to relieve cholinergic symptoms varies and is often unpredictable. There are many atropinization procedures, and no standard regimen has been established [37]. There are two methods that are used. In both cases, atropinization is based on clinical symptoms. Targets for atropinization include areas of acceptable BP, minimum HR, and clear lungs, as low heart rate (HR), low blood pressure (BP), and severe secretions can increase morbidity [38, 39]. The dose of atropine is titrated to achieve these objectives. Some people do not respond to typical atropine doses and require higher doses of atropine to achive atropinization goals. Despite reversal of other cholinergic symptoms such as bronchorea and increased salivation, these patients may have an inadequate heart rate response (70/min) and/or continuously low systolic blood pressure (90 mm Hg). Bradycardia increases the risk of malignant arrhythmias in this situation [40]. Hyperthermia, psychosis, delirium, seizures, or coma can all be caused by high doses of atropine [41, 42].

In a prospective study of patients with carbamate poisoning in the intensive care unit, Samprathi et al., found that 13.6 percent of the patients required atropine and that in these patients, supplemental epinephrine infusion could rapidly achieve the target heart rate without increasing the atropine dose. Patients who required a high dose of atropine were on the ventilator longer and spent more time in the ICU. The group receiving high-dose atropine had a higher mortality rate [43].

Stojiljkovic et al., reported that all the antidotes (atropine, hexamethonium, D-tubocurarine and HI-6) used in their study against carbamate poisoning provided some protection against the lethal effects of carbamates in rats when administered as monotherapy. However, when they used Bispyridinium oxime HI-6 in poisonings with physostigmine and pyridostigmine, they found that the carbamates induced significant reactivation of AChE in the brain and diaphragm, tissues where they act [44].

Non-Pharmacological Treatment Options

The cornerstone of treatment for a poisoned patient is good supportive care. Carbamate poisoning leads to respiratory arrest, which may require intubation or ventilation. Respiratory arrest is a sign of poor prognosis [45]. Elevation of head, prevention of deep venous thrombosis and lung-sparing ventilatory settings should be used with caution in a ventilated patient. There is no significant evidence that hemodialysis or hemoperfusion improves outcomes in carbamate poisoning [46].

Medications for Future Use

Alpha2-adrenergic receptor agonists such as clonidine reduce the synthesis and release of acetylcholine at presynaptic junctions, but their effects in humans are unknown [47].

Huperzine A and ZT-1: Huperzine A is a reversible selective AChE inhibitor and should be co-administered with imidazenil (a non-sedating benzodiazepine) to prevent seizures. Huperzine A acts by blocking NMDA-induced exitatory toxicity and thus has a beneficial effect in preventing seizures and status epilepticus following exposure [48].

Medical Prognosis

The acute physiology and chronic health status assessment II and the Simplified Acute Physiology Score II have been proven to be better predictors of death in acute carbamate poisoning [49]. A Glasgow coma score (GCS) of less than 13 was associated with poor outcome by researchers in a prospective analysis of 1365 patients acutely poisoned with carbamate [50].

Conclusion

Carbamate insecticides are chemicals that cause significant illness and mortality. Due to their widespread use, they can cause occupational, accidental, and intentional poisoning. To reduce morbidity and fatalities, prompt recognition, timely administration of the antidote at appropriate doses and durations, adequate monitoring and supportive care, and proactive poisoning management are essential. Supportive therapy, as well as vigorous and early administration of antimuscarinics and oximes are part of the treatment plan. Although acute carbamate poisoning may resolve within 24 to 48 hours, clinicians should be aware that delayed neuropathy may develop in these patients. Prevention, on the other hand, appears to be the cornerstone of treatment. The use of carbamate compounds in developing countries should be closely controlled by the relevant authorities and the use of natural pesticides instead of chemical pesticides should be encouraged.

References

[1] Morais, S., Dias, E., de Lourdes Pereira, M. (2012). Carbamates: Human exposure and health effects. In: *The Impact of Pesticides.* 1st ed. Academy Publish. Org. 21-38.

[2] Casida, J. E., Augustinsson, K. B., Jonsson G. (1960). *Stability, toxicity and action mechanism with esterases of certain carbamate insecticides. Ibid.,* 53(2), 205-212.

[3] Tintinalli J E, Kelen G D, Stapczynski J S, eds. (2004). *Herbicides and Rodenticides. Emergency Medicine: a Comprehensive Study Guide.* 6th Edn. McGraw-Hill Co, New York, pp. 1134–1143.

[4] Yamashita, M., Yamashita, M., Tanaka, J., Ando, Y. (1997). Human mortality in organophosphate poisonings. *Vet. Hum. Toxicol.*, 39:84-85.

[5] 5. Brahmi, N., Mokline, A., Kouraichi, N., Ghorbel, H., Blel, Y., Thabet, H., Hedhili, A., Amamou, M. (2006). Prognostic value of human erythrocyte acetyl cholinesterase in acute organophosphate poisoning. *Am. J. Emerg. Med.,* 24,822-827.

[6] Eddleston, M., Clark, R. (2011). Insecticides: organophosphorus compounds and carbamates. In: *Nelson LS, editor. Goldfrank's toxicologic emergencies.* New York: McGraw Hill Medical; p. 1450–66.

[7] Gunnell, D., Eddleston, M., Phillips, M. R., Konradsen, F. (2007). The global distribution of fatal pesticide self-poisoning: systematic review. *BMC Public Health*, 7:357.

[8] National Crime Records Bureau. *Accidental deaths and suicides in India - 2010.* Available at: http://ncrb.nic.in/ADSI2010/home.htm. Accessed June 4, 2014.

[9] Mowry J B, Spyker D A, Cantilena L R Jr, Bailey, J. E., Ford, M. (2013). Annual Report of the American Association of Poison Control Centers' National Poison Data System (NPDS): 30th Annual Report. *Clin. Toxicol. (Phila),* 51(10), 949–1229.

[10] Maddy, K. T., Edmiston, S., Richmond, D. (1990). Illness, injuries, and deaths from pesticide exposures in California 1949–1988. *Rev. Environ. Contam. Toxicol.,* 114, 57–123.

[11] Wesseling, C., McConnell, R., Partanen, T., Hogstedt, C. (1997). Agricultural pesticide use in developing countries: health effects and research needs. *Int. J. Health Serv.,* 27(2), 273–308.

[12] King, A. M., Aaron, C. K. (2015). Organophosphate and Carbamate Poisoning. *Emerg. Med. Clin. North Am.,* 33(1), 133-151.

[13] Cumming, G., Harding, L. K., Prowse, K., 1968. Treatment and recovery after massive overdose of physostigmine. *Lancet*, 2, 147–149.

[14] 14. Sener, S., Ozsarac, M. (2006). Case of the month: rivastigmine (Exelon) toxicity with evidence of respiratory depression. *Emerg. Med.,* 23, 82–83.

[15] Hoffman, R. S., Manini, A. F., Russell-Haders, A. L., Felberbaum, M., Mercurio-Zappala, M. (2009). Use of pralidoxime without atropine in rivastigmine (carbamate) toxicity. *Hum. Exp. Toxicol.,* 28, 599–602.

[16] Suzuki, Y., Kamijo, Y., Yoshizawa, S., Fujita, Y., Usui, K., Kishino, T. (2017). Acute cholinergic syndrome in a patient with mild Alzheimer's type dementia who had applied a large number of rivastigmine transdermal patches on her body. *Clin. Toxicol. (Phila),* 55, 1008–1010.

[17] Stojiljković M P, Škrbić R, Jokanović M, Kilibarda V, Bokonjić D, Vulović M. (2018). Efficacy of antidotes and their combinations in the treatment of acute carbamate poisoning in rats. *Toxicology,* 408, 113-124.

[18] Farago, A. (1969). Suicidal, fatal Sevin (1-naphthy1-N-methylcarbamate) poisoning. *Arch. Toxicol.,* 24, 209–215.

[19] Ichikawa, S., Goto, T., Umetsu, N. (1995). Antidotal action of atropine sulfate against insecticide benfucarb poisoning in rats. *J. Toxicol. Sci.,* 20, 143–148.

[20] Brittain, M. K., McGarry, K. G., Moyer, R. A., Babin, M. C., Jett, D. A., Platoff, G. E., Yeung, D. T. (2016). Efficacy of recommended pre-hospital human equivalent doses of atropine and pralidoxime against the toxic effects of carbamate poisoning in the Hartley guinea pig. *Int. J. Toxicol.*, 35, 344–357.

[21] Stojiljković, M. P. (1994). Therapeutic efficacy of antidotes in acute carbaryl poisoning in mice and rats. *Arh. Farm*, 44, 472–473.

[22] Mercurio-Zappala, M., Hack, J. B., Salvador, A., Hoffman, R.S. (2007). Pralidoxime in carbaryl poisoning: an animal model. Hum. *Exp. Toxicol.,* 26, 125–129.

[23] Kalkan, S., Erdogan, A., Aygoren, O., Capar, S., Tuncok, Y. (2003). Pesticide poisonings reported to Drug and Poison Information Center (DPIC) in Izmir, Turkey. *Vet. Hum. Toxicol.,* 45, 50-52.

[24] Perola, E., Cellai, L., Lamba, D., Filocamo, L., Brufani, M. (1997). Long chain analogs of physostigmine as potential drugs for Alzheimer's disease: new insights into the mechanism of action in the inhibition of acetylcholinesterase. *Biochim. Biophys. Acta.,* 1343, 41–50.

[25] Jokanović, M. (2009). Medical treatment of acute poisoning with organophosphorus and carbamate pesticides. *Toxicol. Lett.,* 190, 107–115.

[26] 26. Kahraman N. (2008). Organofosfat ve karbamat içeren insektisid maruziyeti ile acil servise başvuran hastaların serum asetilkolinesteraz düzeyleri ile klinik seyir ve mortaliteleri arasındaki ilişkinin değerlendirilmesi. Uzmanlık Tezi, İzmir, Türkiye.

[27] Pinakini, K. S., Kumar, T. S. M. (2006). Serial cholinesterase estimation in carbamate poisoning. *J. Clin. Forensic Med.,* 13, 274–276.

[28] Kahraman, N., Yanturalı, S., Kalkan, F., Oray, N. F., Hocaoğlu, N., Uğurhan A. (2008). Organofosfat ve Karbamat İçeren İnsektisid Zehirlenmelerinde Serum Asetilkolinesteraz Düzeyleri ile Klinik Seyir ve Mortalite Arasındaki İlişkinin Değerlendirilmesi. *Turk. J. Emerg. Med.,* 8(3):121-126.

[29] Robey, W. C., Meggs, W. J. (2004). Insecticides, Herbicides and Rodenticides. In: Tintinalli, J. E., Kelen, G. D., Stapczynski, J. S., eds. *Emergency Medicine: a Comprehensive Study Guide.* 6th Edn. McGraw-Hill Co, New York, 2004; pp. 1134–43.

[30] Worek, F., Kirchner, T., Backer, M., Szinicz, L. (1996). Reactivation by various oximes of human erythrocyte acetylcholinesterase inhibited by different organophosphorus compounds. *Arch. Toxicol.,* 70, 497–503.

[31] Karki, P., Ansari, J. A., Bhandary, S., Koirala, S. (2004). Cardiac and electrocardiographical manifestations of acute organophosphate poisoning. *Singapore Med. J.,* 45, 385-389.

[32] Eyer, P. (2003). The role of oximes in the management of organophosphorus pesticide poisoning. *Toxicol. Rev.,* 22, 165–90.

[33] Murray, J. C., Stein, F., McGlothlin, J. C. (1994). Prolongation of the prothrombin time after organophosphate poisoning. *Pediatr. Emerg. Care.,* 10, 289-290.

[34] Tuncok, Y., Aksay Hocaoğlu, N. (2006). Organofosfatlı İnsektisidlerle Zehirlenme, Turkiye Klinikleri. *J. Surg. Med. Sci.,* 2, 69-73.

[35] Stojiljković, M. P., Sjauš, T. K., Dogović, N. (1989). Relative therapeutic efficacy of anticholinergic drugs in the treatment of acute physostigmine intoxication in rats. *Iugoslav. Physiol. Pharmacol. Acta.,* 25, 139–140.

[36] Chyka, P. A., Seger, D. (1997). Position statement: single-dose activated charcoal. American Academy of Clinical Toxicology; European Association of Poisons Centres and Clinical Toxicologists. *J. Toxicol. Clin. Toxicol.,* 35, 721-741.

[37] Eddleston, M., Buckley, N. A., Eyer, P., Dawson, A.H. (2008). Management of acute organophosphorus pesticide poisoning. *Lancet.* 371(9612), 597–607.

[38] Eddleston, M., Buckley, N. A., Checketts, H., Senarathna, L., Mohamed, F., Sheriff, M. H. R., Dawson, A. (2004). Speed of initial atropinisation in significant organophosphorus pesticide poisoning a systematic comparison of recommended regimens. *J. Toxicol. Clin. Toxicol.,* 42, 865–875.

[39] Perera, P. M., Shahmy, S., Gawarammana, I., Dawson A. H. (2008). Comparison of two commonly practiced atropinization regimens in acute organophosphorus and carbamate poisoning, doubling doses vs. ad hoc: a prospective observational study. *Hum. Exp. Toxicol.,* 27, 513–518.

[40] Solti, F. (1989). Malignus aritmiak [Malignant arrhythmia]. *Orv. Hetil.,* 30, 163–166.

[41] O'Connor, P. S., Mumma, J. V. (1985). Atropine toxicity. *Am. J. Ophthalmol.,* 99, 613–614.

[42] Soulaidopoulos, S., Sinakos, E., Dimopoulou, D., Vettas, C., Cholongitas, E., Garyfallos, A. (2017). Anticholinergic syndrome induced by toxic plants. *World J. Emerg. Med.,* 8, 297–301.

[43] Samprathi, A., Chacko, B., Reynal D'sa, A., Rebekah, C. G., Kumar, V., Sadiq, M., Victor, P., Prasad, J., Jeevan Jayakaran, J. A., Peter, J. V. (2021). Adrenaline is effective in reversing the inadequate heart rate response in atropine treated organophosphorus and carbamate poisoning. *Clin. Toxicol. (Phila),* 59(7), 604-610.

[44] Stojiljković, M. P., Škrbić, R., Jokanović, M., Kilibarda, V., Bokonjić, D., Vulović, M. (2018). Efficacy of antidotes and their combinations in the treatment of acute carbamate poisoning in rats. *Toxicology,* 408, 113-124.

[45] Noshad, H., Ansarin, K., Ardalan, M. R., Ghaffari, A. R., Safa, J., Nezami, N. (2007). Respiratory failure in organophosphate insecticide poisoning. *Saudi Med. J.,* 28(3), 405–407.

[46] Altintop, L., Aygun, D., Sahin, H., Doganay, Z., Guven, H., Bek, Y., Akpolat, T. (2005). In acute organophosphate poisoning, the efficacy of hemoperfusion on clinical status and mortality. *J. Intensive Care Med.,* 20(6), 346–350.

[47] Balali-Mood M., Saber, H. (2012). Recent advances in the treatment of organophosphorous poisonings. *Iranian J. Med. Sci.,* 37, 74.

[48] Kumar, S. V., Fareedullah, M. D., Sudhakar, Y., Venkateswarlu, B., Kumar, E.A. (2010). Current review on organophosphorus poisoning. *Arch. Appl. Sci. Res.,* 2(4), 199–215.

[49] Peter, J. V., Thomas, L., Graham, P. L., Moran, J. L., Abhilash, K. P., Jasmine, S., Iyyadurai, R. (2013). Performance of clinical scoring systems in acute organophosphate poisoning. *Clin. Toxicol.*, 51, 850–854.

[50] Narang, U., Narang, P., Gupta, O. (2015). Organophosphorus poisoning: a social calamity. *J. Mahatma Gandhi Institute Med. Sci.,* 20, 46.

About the Editor

Güllü Kaymak completed her bachelor's degree in Biology at Ondokuz Mayıs University in 2008, her master's degree in Environmental Science at Marmara University in 2011, and her doctorate degree in Biology at Sakarya University in 2017. Then, she accomplished her postdoc project at Kütahya Health Sciences University with a National Postdoc Scholarship from TUBITAK (The Scientific and Technological Research Council of Turkey). Now, she is currently working at Kütahya Health Sciences University as a lecturer. She plans to continue her scientific studies in the fields of toxicology, especially aquatic ecotoxicology, and environmental pollution. She aspires to study on building up new strategies for preventing the accumulation of pollutants and the remediation of polluted ecosystems. She believes her current studies will enrich the knowledge about the effects of carbamates on the environment and organisms. Thus, she hopes it will provide new strategies to protect and sustain a healthy ecosystem.

List of Contributors

Elif Aydın, M.Sc.
Disinfection, Sterilization and Antisepsis Program,
Kütahya Health Sciences University, Kütahya, Turkey.

Rahul Badru, PhD
Department of Chemistry, Sri Guru Granth Sahib World University,
Fatehgarh Sahib, Punjab, India.

Khushboo Verma
Department of Chemistry, Sri Guru Granth Sahib World University,
Fatehgarh Sahib, Punjab, India.

Derya Berikten, PhD
Training and Research Center, Kütahya Health Sciences University,
Kütahya, Turkey.

Yalçın Dicle, PhD
Department of Nutrition and Dietetics, Muş Alparslan University,
Muş, Turkey.

Serdal Öğüt, Prof. Dr.
Aydın Adnan Menderes University, Faculty of Health Sciences,
Aydın, Turkey.

Meliha Koldemir-Gündüz, PhD
Department of Engineering Basic Sciences, Kütahya Health Sciences
University, Kütahya, Turkey.

Index

R

S

T

U

V

X